Magnetic Light

Discovering
the
Two Independent Components
of
Electromagnetic Waves

Duncan Crane

First Printing, 2020

ISBN 978-0-578-72281-8

Dedicated to Barbara, the love of my life.
Who, though she speaks only a few words of Science,
has always supported my interest in how things work
with true heartfelt eloquence.
Thank you.

"The world is so full of a number of things,
I'm sure we should all be as happy as kings."

Robert Louis Stevenson

Contents

Introduction 1
1 - Electromagnetic Waves 3
2 - Original Data 7
3 - Presentation 13
4 - New Data 15
5 - The Math 27
6 - Speed of the Electric Wave 30
7 - Full Accounting for Phase Shift 38
8 - Oleg Dmitrovitch Jefimenko 41
9 - Additional Thoughts 43
10 - Conclusion 48
Appendix 1: Schematic for Dad's Radios 50
Appendix 2: Schematic for My Radios 52
Appendix 3: Loopstick Design 54
Appendix 4: Wraparound 57
Appendix 5: Data 60
Appendix 6: Data – Mobile 78
Appendix 7: Phase Shift Script 82
Appendix 8: Authors 85
References 88

Introduction

In 2001 I attended the wedding of my sister's son at a resort near Portland, Oregon. My dad was there too and after the reception we slipped over to the pub to discuss what our next science project would be. He was a retired optical physicist and I was a computer programmer twelve years out from retirement. We had been doing science projects of one kind or another for virtually all my life. Oscilloscope on a cake pan, radar in an aluminum flying saucer sled, laser interferometer water tunnel, laser-based speedometer, deep sea salinity gauge. We were currently in between projects and needed to decide what next.

Dad asked me what I was interested in. I thought for a moment and said, "You know, I have never really understood how the two components of light work together during propagation, the electric and magnetic." He thought that over and said, "Yes. That has always bothered me too." We agreed right then and there that "Light" would be our next project. We shook hands on it and Dad saved the coaster from his beverage as a memento of the event.

Five years later Dad had filled four large binders with notes on 14 different experiments all directed at a better understanding of light. He called himself an experimental physicist as opposed to a theoretical physicist, which meant that he had to see everything with his own eyes and come to a deep intuitive grasp of how it worked. He took very little on faith. These 14 experiments started with static electric and static magnetic fields and built up to the full electromagnetic theory of light that we all know and love as "Maxell's Equations". [Ref 1]

One of the things my dad wanted to see for himself was the relationship between the electric and magnetic fields while propagating. Did they really travel side by side, one vertical and the other horizontal, in perfect step with one another? This he covered in experiment Number 7. This book focuses on that experiment and what he and I discovered while doing it.

1 - Electromagnetic Waves

Sunlight, starlight, radio waves, cell phone waves and the waves in our kitchen microwaves are all electromagnetic waves. What do they look like?

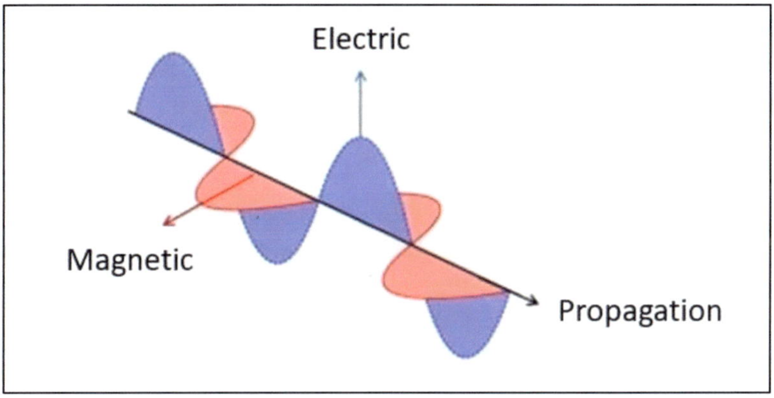

Figure 1: Electromagnetic wave [Ref 2]

Figure 1 is the standard depiction of an electromagnetic wave. The electric field goes up and down while the magnetic field goes side to side. When the electric field is at its peak in the upward direction, the magnetic field is at its peak in the leftward direction. When the electric is at its peak in the downward direction, the magnetic field is at its peak in the rightward direction. The two fields ebb and flow in directions that are at right angles to one another, the vertical and horizontal directions, and they do so with perfectly synchronized timing. This is what is meant when we say that the electric and magnetic fields of electromagnetic waves are "in phase". I like to say that the two fields are "joined at the hip".

It is Maxwell's famous equations that gave us the understanding for this diagram. A changing electric field will create a magnetic field; and a changing magnetic field will create an electric field. So one field creates the other field while the other is creating the one. This is where my head starts to spin and why I never really understood light. Traditionally when a student in the classroom raises their hand to ask for clarification on this point they are told that it is indeed hard to visualize but it must be correct because Maxwell was able to derive the all-important constant c_o (the speed of light in a vacuum) by feeding the fundamental nature of both electric and magnetic fields into the standard equation for the speed of waves of any kind. These two fields must be sustaining one another somehow, cycle after cycle. To do this of course they would need to be in phase.

The best way to see an electromagnetic wave has always been to display it on an oscilloscope. Hang a wire out the window and attach it to an oscilloscope and you will likely be looking at the signal from the nearest and strongest AM radio station in your neighborhood. It might look like Figure 2.

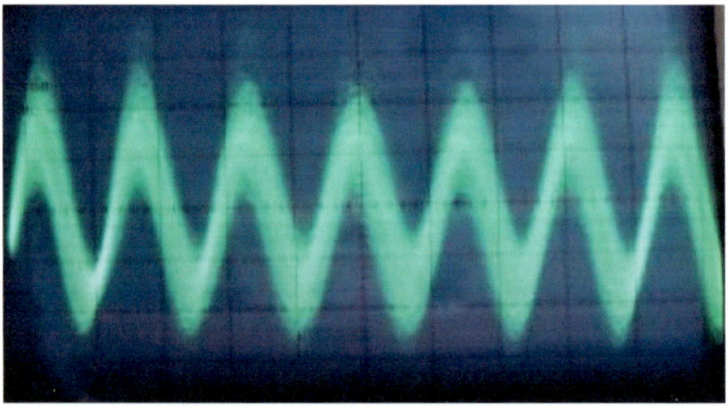

Figure 2: WNNZ on analog oscilloscope

This is what I get on my old analog scope when I attach five feet of wire to channel 1. I have to increase the sweep speed to get the wave to emerge, but after doing the calculations with sweep speed and hash marks per cycle I can see that this is WNNZ, the

50,000 Watt AM station transmitting at 640 KHz from its transmitter just six miles down the road.

The picture is fuzzy because the audio riding on the 640 KHz carrier is causing the amplitude to jump up and down, as it should. This is how AM (amplitude modulation) radio works. If I turn on my AM radio and tune in WNNZ I can easily see that the wave on the oscilloscope is "jiggling" in sync with what I am hearing – confirming that I am looking at the signal from WNNZ.

When the electric field of the station's signal hits my five foot wire antenna it generates an electric voltage in the wire that is visible on the scope. The magnetic component of the station's signal is just as simple to pick up but a loop antenna must be used instead of a straight wire.

Figure 3: Loop antenna, 18 in diameter, 26 turns of AGW 20 copper wire

Loop antennas need to be rotated from side to side to find the maximum signal strength. When the opening of the loop is oriented at 90 degrees to the station's transmitter, the maximum amount of magnetic field can sweep back and forth through the open loop, generating a voltage in the wires of the loop which can then be displayed on the scope.

On the other hand, when the loop is oriented with one of its open faces aimed directly toward the transmitter, there will be zero magnetic signal picked up because none of the magnetic field can sweep through the open loop.

When displayed on the scope, the magnetic signal looks just like the electric. But if you display one directly on top of the other, you can see that the E signal is arriving AFTER the M signal.

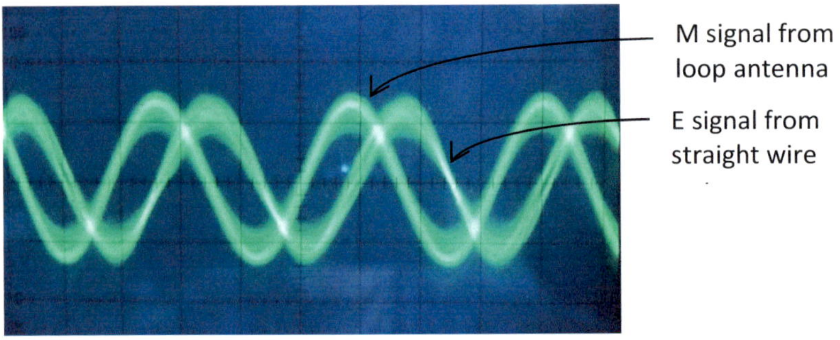

Figure 4: Magnetic and electric signals from WNNZ

If you have a wire antenna, a loop antenna and a dual trace scope, you can see for yourself if the electric and magnetic components of an electromagnetic wave are in phase. There are issues such as careful tuning and inherent phase shift in the equipment that must be considered but basically this sums up my dad's experiment Number 7.

2 - Original Data

The first time my dad showed me the results from experiment Number 7, I had serious doubts. I looked at his setup and thought how there were so many ways for extraneous phase shift to creep in. How could he be so sure of his results? I knew how sensitive a wire antenna was. Touching it or just standing nearby can change everything. The loop antenna is just as sensitive and has the added factor of depending on the direction it is pointed. Everything inside the radio and many things outside generate "stray capacitance". Every pico (10^{-12}) farad of stray capacitance changes the tuning and ultimately the measured phase shift.

My dad had a radio for each antenna with two stages of tuning in each. A simple radio has one stage of tuning but if you want good selectivity you need two stages. However, two stages of tuning are famous for being difficult to work with. It is like the difference between riding a unicycle, and balancing a twirling plate on a stick while riding a unicycle. And tuning would have to be near perfect or the phase shift we were looking for would not appear.

The transistors and other components that made up the two radios would each introduce a bit of phase shift that had nothing to do with the phase shift in question. I was impressed with the results but knew that the only way for me to really believe the results would be to build my own setup from the ground up. I had the benefit of being trained as a ground radio repairman by the US Air Force but that was many years earlier. I would have to teach myself current radio technology.

Figure 5 below is a simple block diagram of how my dad used two homemade radio receivers and two antennas (straight wire and loop) to capture the magnetic and electric signals of a given AM station and display those signals on an oscilloscope and thereby see and measure the difference in their arrival times. He first used this setup in 2005.

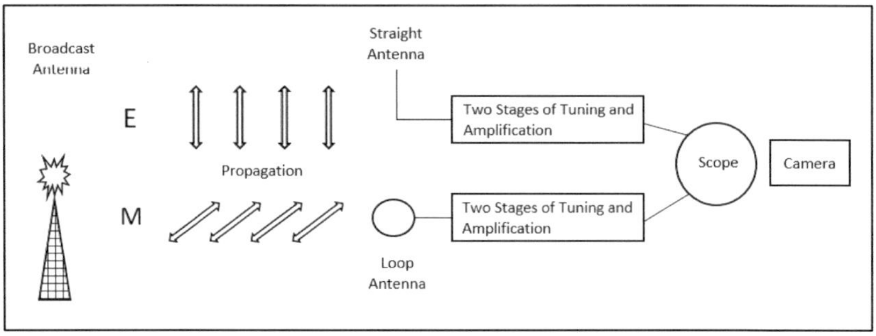

Figure 5: Block diagram for Dad's setup

Figure 6: Photograph of Dad's full setup in his back yard in Southbury CT where he took many of his measurements

Below is a picture of what he typically saw on his analog scope when he compared the electric and magnetic signals. Again, the signals are blurry because of the audio riding on the carrier. But as my dad pointed out, if you look closely at where the signals cross the zero line, they can become quite sharp and give good phase shift measurements.

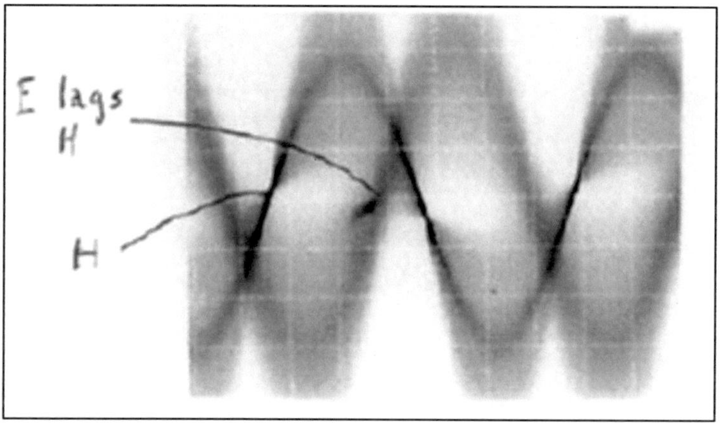

Figure 7: Magnetic and electric signals on Dad's oscilloscope

As Dad was a trained physicist he used the symbol "H" (and sometimes B) for the magnetic field while I am simply using "M".

On the next page is a record of phase shift readings for five different stations taken 11/2/2005 by my dad at his home in Southbury, CT. He recorded his data by hand and I have transcribed it here to a spreadsheet to get the linear regression line that connects the data points.

Robert Crane Jr, 11/2/2005 , Southbury CT

Electric component appears to fall behind the magnetic component at a rate of roughly 10 degrees per mile.

Station	Location	Miles	Freq	Shift (degrees)
WINE	Danbury CT	12	940 KHz	221
WELI	New Haven CT	23	960 KHz	319
WTIC	Hartford CT	41	1080 KHz	528
WINS	NYC	62	1010 KHz	720
WCBS	NYC	62	880 KHz	724

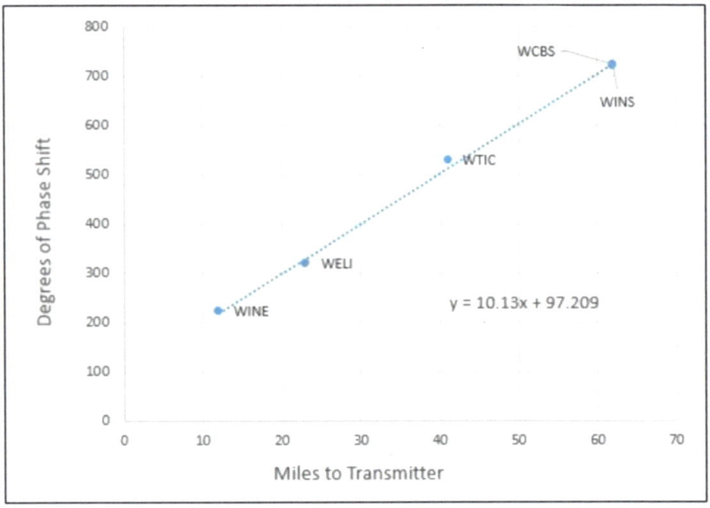

Figure 8: Data recorded by Robert Crane, 11/2/2005, Southbury CT

Notice that the five stations are listed in order of distance which is also in order of increasing phase shift. On the graph the five data points fall along a reasonably straight line whose slope is 10.13 degrees of phase shift per mile to the transmitter.

As mentioned earlier, conventional wisdom states that the E (electric) and M (magnetic) components of an electromagnetic wave stay strictly in sync with one another from a timing perspective. If they were not there to support the creation of one another cycle after cycle, the wave would cease to propagate and cease to exist. But here we have the signal coming in on the E radio falling further and further behind the signal coming in on the M radio. It's a horse race! The E horse is slower than the M horse and the longer they race the further the E horse falls behind. This is the data that first revealed that these two components are not "joined at the hip".

Now, divide the measured phase shift between the E and the M signals by the miles between transmitter and receivers and come up with a phase shift per mile figure. Surprisingly, this phase shift per mile rate is the same for all five stations shown above on my dad's data. This is why the five stations fall so nicely on a straight line. They all have a different measured phase shift because the signals have traveled different distances, but when their phase shifts are divided by their respective distances they share a common phase shift per mile rate.

However, a given station's phase shift measurement is based upon its wavelength which of course gets longer as frequency decreases. So the slower speed of the E signal is not a flat rate like the M (which we are assuming remains at the speed of light after it leaves the broadcast tower) but varies by frequency, dropping as the frequency drops at just the right rate so as to maintain a common phase shift per mile across all AM band frequencies. I found this very surprising at first. (More on this later.)

There was a difference in Dad's setup when compared to mine. It is not clear what that difference was but it introduced an addition 90 degrees of inherent phase shift in all of his readings. You can see this in the graph of his 11/2/2005 data. If you follow the line all the way to the left (zero miles) you can see that the line starts with roughly 90 degrees of phase shift at its origin. This was most likely due to a wiring difference in our radios. The critical result is that the shift per mile was 10 degrees for both of our setups.

There is one more thing worth pointing out in this first sample of data. Notice that at roughly mile 27 on the horizontal axis the corresponding degrees on the vertical axis is roughly 360 degrees. Initially when my dad was recording these measurements it did not occur to him that phase shifts above 360 would be possible. So the stations that were beyond 27 miles had phase shift readings that placed them down among the other closer stations on the graph. Looking at this collection of data points, he could see no discernible pattern - yet obviously something was going on.

Then out of the blue a few days later the idea of "wrap-around" came to him. He called me in a very excited state and just said "Wrap-around"! Thereafter whenever the known location of a given transmitter was greater than 27 miles he would add 360 degrees to the measured phase shift before calculating the degrees per mile for that station. If the station was an additional 36 miles beyond that he would add another 360 degrees. This put all the stations on a steady 10 degree per mile line. (See Appendix 4 for more on wraparound.)

3 - Presentation

By 2009 my dad felt it was time to tell the world what we had discovered. He wrote up a paper and submitted it to the International Society for Optics and Photonics (SPIE). The paper was accepted for presentation at SPIE's August 2-6, 2009 conference in San Diego, CA. We discussed what the trip would entail and decided that it would be best if I were to present the paper.

SPIE. The international society for optics and photonics

The conference was held at the San Diego Convention Center and was a very big deal. There were thousands of attendees from all over the world and hundreds and hundreds of papers and exhibits. It was very exciting. Dad and I were a little bit surprised that his paper was accepted because its basic premise was not something most physicists would take seriously. The title of his paper was "Light's Two-Fields". He felt that maybe SPIE gave him the benefit of the doubt and accepted the paper out of courtesy since he had authored a number of respected papers in the past and held a number of patents in the field of optics.

LIGHTS TWO-FIELDS

Robert Crane
Consultant, Southbury, CT 06488, USA

ABSTRACT

This paper offers a "Two-Field" point of view for understanding propagating electromagnetic energy. Nature's parsimony allows us to say, "Light beams consist of two continuous, independent, modulated and rapidly moving E and H fields." At lower intensities the intermittent emission identified by Planck appears. Experiments involving radio fields showed that E fields propagate at slower velocity than H fields. Experiments involving light showed that H fields passing through the atmosphere will probably not be scintillated. Experiments are underway to photograph scenes using the H portion of light. Experiments are underway to reduce blur of an over-exposed diffraction-limited image.

Figure 9: Abstract from Robert Crane's SPIE paper, 8/2009 [Ref 3]

At the conference I presented the paper to a small group of roughly thirty scientists and engineers. Our conference track title was "Optical Engineering and Applications". I felt seriously outclassed. Maxwell's equations were being tossed about like they were everyone's standard diet. When it was my turn I did what I thought was a passable job of presenting our data and looked forward to the question and answer period at the end. But there were no questions. No comments. Just silence. That said it all – the idea of the electric and magnetic fields traveling independently was unthinkable.

4 - New Data

My dad passed away in 2017 at the age of 97. On the way home from his memorial I decided that I needed to get serious about reproducing the results of experiment Number 7. Three years later I am fully convinced that the E and M fields travel separately.

I had to train myself in transistor electronics. I had to learn the peculiar properties of loop and loopstick antennas. I wrestled with the Q (selectivity) of my radios. Input impedance. The critical nature of tuning. The jungle of AM stations in my area with what felt like overlapping frequencies and stations whose frequencies are harmonics of one another. I found that averaging could hide a lot of the “jiggles” that the audio gave to the carrier wave. I identified a number of sources of interference that covered up what I was looking for. Ultimately I got my setup working right and could clearly see the 10 degrees per mile phase shift seen by my dad.

Here is a block diagram of my set up, much like my dad’s.

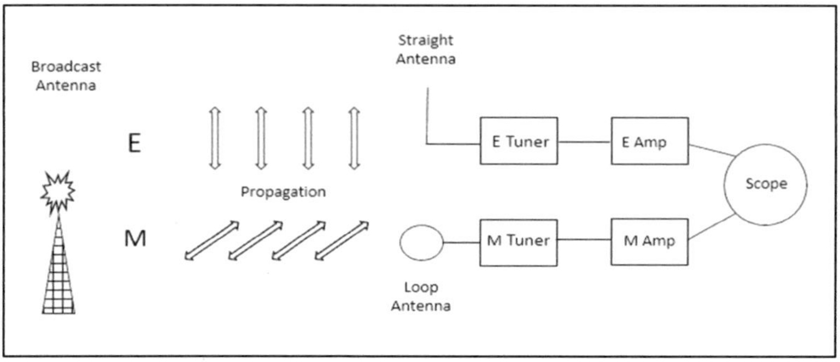

Figure 10: Block diagram for my setup

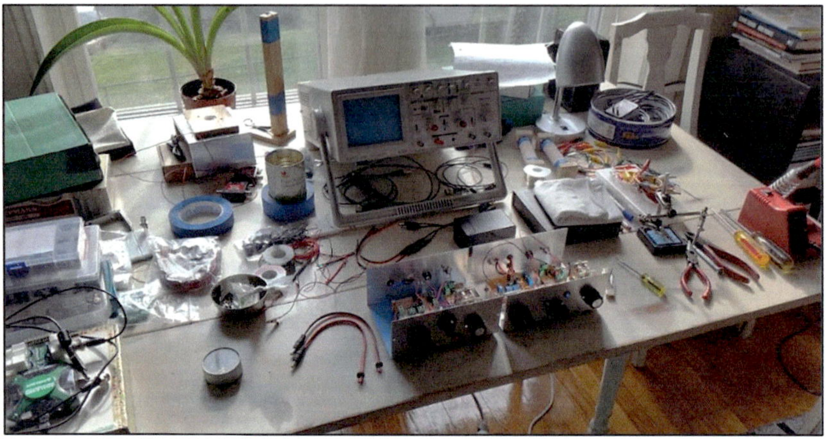

Figure 11: Picture of my set up under construction

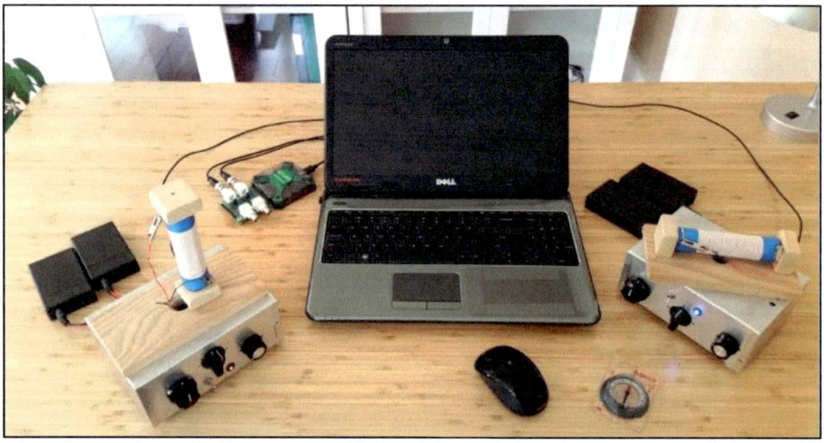

Figure 12: Picture of my setup in its final configuration

I made two big changes in my setup compared to my dad's. First I replaced the analog oscilloscope with a digital oscilloscope, with the inputs coming from a powerful two channel analog-to-digital sampler made by Digilent. [Ref 4] The sampler grabs the profiles of the electric and magnetic signals as they exit their respective radios and samples their voltage level at 100 million times a second.

Second, I replaced the standard straight wire and loop antennas with two custom designed loopsticks. A loopstick is just a loop antenna with a ferrite iron core. (I describe the design of these loopsticks in Appendix 3.) Using a loopstick instead of a loop antenna makes sense for the magnetic signal but how could I use a "magnetic antenna" for the electric signal? I stood it up in a vertical position. In that orientation it worked fine as an electric antenna because no horizontally sweeping magnetic fields could pass through its core. It is basically acting as nothing more than a wire.

The value of using matching loopsticks as antennas on both radios was that it gave me one design to use for both radios, including antennas. Identical designs and construction went a long way towards eliminating inherent phase shift in my equipment. The loopsticks themselves have two functions to perform in each radio: antenna and tuning tank coil.

Notice the little lights on each radio - a red light on the electric radio and a blue light on the magnetic. On the oscilloscope the electric signal is displayed in red and the magnetic in blue.

Also visible in the photo above are the 12 Volt battery packs, two for each radio. With the radios being battery driven and the laptop running on its battery the whole setup becomes portable. This was important for road trips. On road trips I started 25 or 30 miles out from a station with a strong signal. I would take a reading, then drive three to five miles towards the transmitter and take another reading, repeating until I reached the transmitter. It is reassuring to see the 10 degrees per mile rule hold for a single station while distance changes. (See Appendix 6 - Data – Mobile.)

Below is a typical picture of what I would see on the digital scope. This particular AM station is 4.6 miles away and the phase shift is measuring 52 degrees so the phase shift per mile comes to about 11 degrees. Throughout this book I refer to the phase shift per mile rate as a standard 10° per mile. In practice, however, this would vary a few degrees either way.

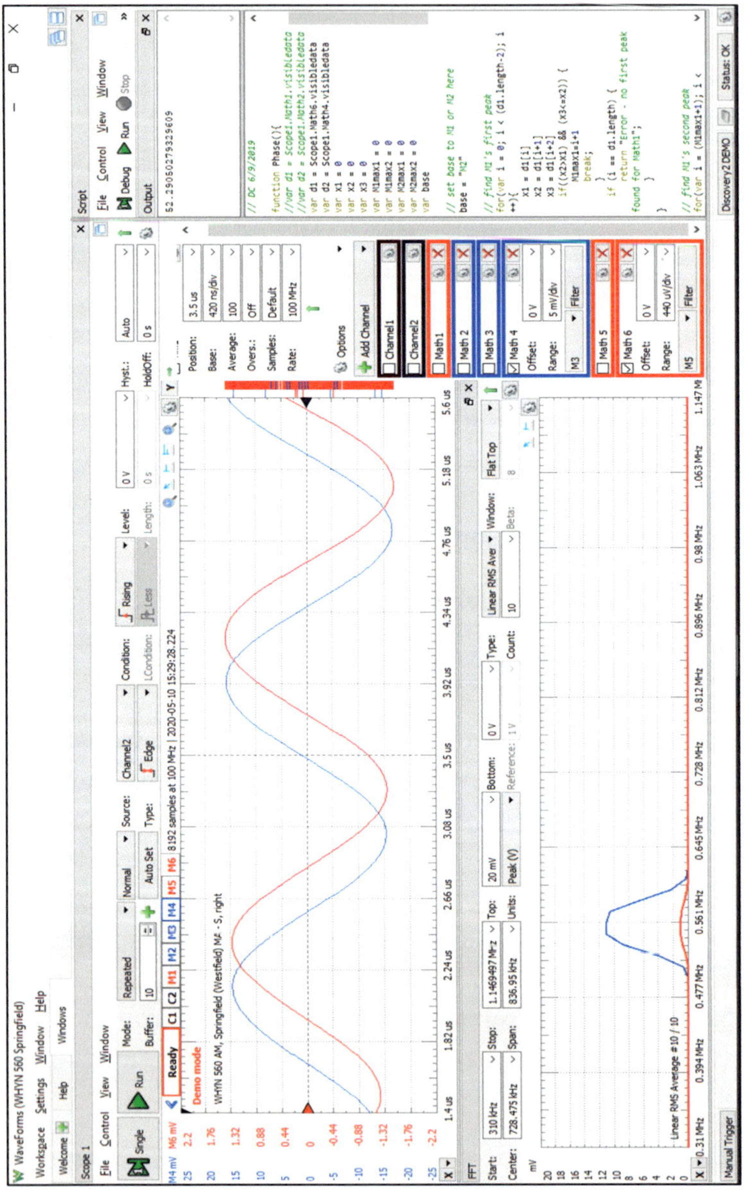

Figure 13: A typical phase shift measurement on digital oscilloscope

The panel on the lower left is the FFT (Fast Fourier Transform) display of signal strength across frequency. The red (the electric radio) is a lot weaker than the blue (the magnetic radio) but both are peaking on the station of interest which is WHYN at 560 KHz.

The FFT screen is the best place to focus when first tuning in a station. Averaging of the signal over 10 samples helps to reduce the audio jiggling. Both radios have a rough tuning variable capacitor (365 pico Farads over one half turn) and a fine tuning variable capacitor (7 pico Farads over 6 turns). At the low end of the AM dial the fine tuner is basically nonresponsive while the rough tuner does the job. At the high end of the AM dial the rough tuner is too responsive to find the precise peak but the fine tuner gets the job done.

When tuning the magnetic radio, the antenna coil must be rotated to get the maximum signal. As it is turned through a full 360 degrees there will be two maximums, both with the open faces of the coil (in my case the two wooden supports at the ends of the loopsticks) at right angles to the direction of the station's transmitter. (Figure 14) If by chance I choose the right maximum, the resulting measured phase shift will be correct. If I choose the wrong maximum, the measured phase shift will be off by 180 degrees.

This is because the direction of the magnetic waves sweeping through the loopstick determine which way the induced current will flow. The two wires that come out of the magnetic radio and attach to the coil are red and black representing positive and ground. Once I determine which maximum is correct I record the fact in the column labeled "Red Wire on M Coil (outbound)" on the data sheet. In doing this I am visualizing the signal outbound from the transmitter and arriving at the coil where the red wire is either on its left or its right.

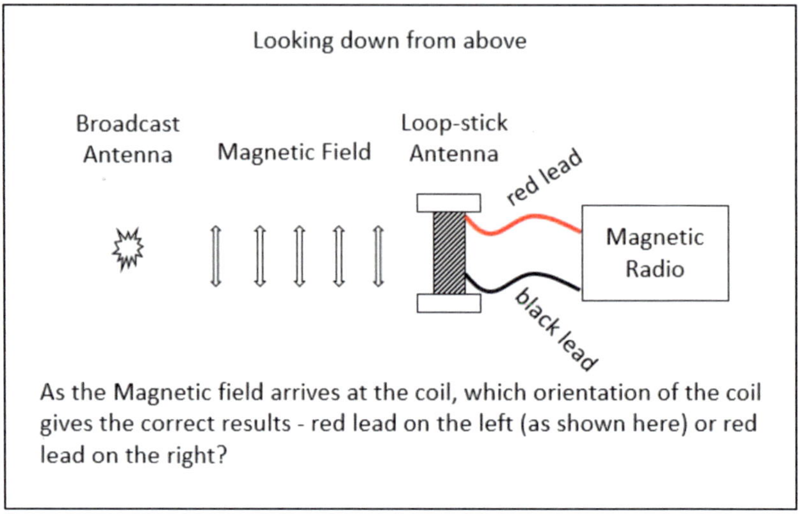

Figure 14: Diagram of how to record orientation of magnetic antenna

Apparently there is no standard for which way AM broadcasting transmitters are wired up in this regard. I have found left or right orientations are roughly 50/50.

With both radios tuned to peak signal strength the focus changes to the Scope Window, upper left on the screen. I turn on averaging here as well (100). At this point the two signals should be refreshing every second or two with a very steady phase shift.

If the signals are steady at this point I do one more round of fine tuning. This time the focus is on the peaks of the waves in the Scope Window instead of the FFT Window. With the trigger set on the magnetic channel, fine tune the electric. While tuning the peak will rise and fall slightly and at the same time travel forward and backward just a little. Tune for the peak. Then set the trigger on the electric and fine tune the magnetic.

Careful tuning is perhaps the most critical aspect of these measurements. If the true peak of both waves cannot be found there is no hope for measuring the phase shift of the electric and magnetic signals. If one or the other signal is slightly off peak, it introduces a phase shift that has nothing to do with the phase shift caused by the different speeds of the two waves. This unwanted phase shift can be thought of as inherent in the equipment at that point in time and must be eliminated.

With fine tuning done and with steady signals, the phase shift between them can be measured by counting grid squares on the scope and doing the math. Or, the tools provided in the upper right corner of the Scope Window can be used to measure the time between a magnetic peak and the electric peak immediately to its right, and the time between two successive magnetic peaks:

> (360) x (time between neighboring M and E peaks) / (time between two successive M peaks).

Even better, the Java Script code in the right hand panel can be run and it will calculate the phase shift. (see Appendix 7)

As just stated, if the gap between the blue wave and the red wave is not reasonably steady no reading can be made. There are two basic reasons why the signal may not be steady - noise and nearby stations,

I have found that the power adapter on my laptop computer generates a lot of radio wave noise. If I forget and leave it plugged in I can drive myself crazy wondering why the signals are not steady. So the laptop must be run on battery while taking measurements. The laptop itself also seems to generate a lot of noise at a couple of spots on the AM dial. I find that placing the two radios three to four feet from the laptop resolves this issue.

The second consideration is nearby stations. Nearby in the sense of being right next to the station of interest in regards to frequency, or for the directional magnetic antenna, in the same or opposite direction as the station of interest. And when a station's frequency is half or twice the frequency of the station of interest it may come through when you think it would not.

Worst of all is the 800 lb. gorilla - 50,000 Watt stations. I have one down the road roughly 6 miles away and it is always inviting itself to the party! Thankfully it drops to 1,000 Watts after sunset allowing me to tune in some stations that otherwise I could not. Adding digital filters via the sampler's software can help eliminate some but not all interfering stations. Just remember, "If it is not steady, it is not ready!"

The ability of a radio's tuner to select a given station and exclude others is called Q. A radio with a high Q does a good job of selecting one station while one with a low Q may pick up more than one station.

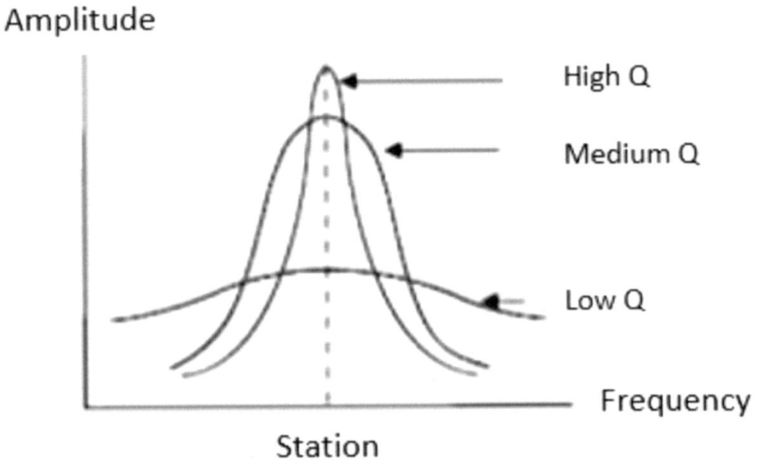

Figure 15: Radio receiver selectivity – Q [Ref 5]

Over the years radios have evolved to have higher and higher Q. The first radios had two or three stages of simple tuning and amplification. Then came regenerative designs that fed a bit of the tuned signal back into the input as positive feedback. Next came superheterodyne radios that had an internal oscillator that tracked with the tuner so that the output was always a fixed 455 KHz signal no matter which station, allowing a tighter design on the Q because the design did not have to deal with the full AM band. Now we have digital radios that sample the signal and feed it through FFT logic to pull out the desired frequency. These have very good selectivity.

As much as I coveted the wonderful selectivity of these more recent designs I did not trust that they would preserve the phase shift between the transmitted electric and magnetic signals. If a radio is designed to pull the audio signal off a carrier whose frequency is roughly 1,000 times that of the audio, who is going to notice if a quarter cycle of carrier goes missing in the process? It is all about the audio. I worried that the true phase shift between the two signals, if it existed, could be totally scrambled by these designs. Thus my decision to stick with the old and simple design of tuning tank plus amplifier.

On the next two pages is a data sheet from a typical day's measurements. Here are a few notes on how to read the data sheet.

- In the upper right corner is the location from which the readings were taken.
- In the left corner the weather is included because at some point that may be useful in understanding the phase shift.
- Wraparound is the distance in miles at which 360 degrees of phase shift is first reached. It is calculated by dividing the average phase shift (cell in lower right) into 360 degrees.
- The basic parameters of the station such as call sign, frequency, city, latitude, longitude, direction, power and distance are self-explanatory.
- Distance is calculated from Lat and Long of station transmitter to Lat and Long of my home, or where I am taking the readings if mobile.
- Red Wire on M Coil indicates the orientation of the M antenna coil.

- Measured Phase Shift is the actual M to E phase shift as measured on the scope.
- Wraparound is calculated by dividing the Distance to the transmitter in miles times 10 by 360, then dropping the remainder. (D*10)/360
- Initial Offset was added for the peculiar behavior of one station: WNNZ. For some reason WNNZ's phase shift measurements acquired an additional 90° of phase shift sometime between 6/2019 and 5/2020.
- Total Shift is the measured Phase Shift plus Wraparound times 360.
- Shift/Mile is the Total Shift divided by the Distance to the transmitter.

The primary piece of information on the data sheets is of course the way the graph on page two of the data sheet has all the stations falling along a straight line starting at the origin and climbing at 10 degrees per mile.

Datasheet - recorded 06/20/2020, page 1					
Easthampton, MA					
Latitude			Longitude		
42	15	39.1	72	40	59.8
42° 15' 39.1'' N			72° 40' 59.8'' W		

Start Time	6:45 AM	
Stop Time	8:30 AM	
Temp	66	F
Humidity	90	%
Ground	dry	dry/wet/snow
Air	dry	dry/rain/snow
Pressure	30.05	inHg
Wrap-around	32.8	mi

	Station	Freq (KHz)	City	Lat			Long			Direction	Power Day (Watts)	Power Night (Watts)	Distance (mi)	Red Wire on M Coil, outbound	Measured Phase Shift (degrees)	Wrap-arounds	Initial Offset (degree)	Total Phase Shift (degree)	Phase Shift / Mile (degree)
1	WMCA	570 AM	NYC	40	45	10	74	6	15	SSW	5,000	5,000	127.7	left	240	3	0	1,320	10.3
2	WEZE	590 AM	Boston, MA	42	24	24	71	5	14	E	5,000	5,000	82.2	right	128	2	0	848	10.3
3	WICC	600 AM	Bridgeport CT	41	9	36	73	9	43	SSW	1,000	500	80.1	right	134	2	0	854	10.7
4	WNNZ	640 AM	Westfield, MA	42	10	44	72	45	4	SSW	50,000	1,000	6.6	right	160	0	90	70	10.5
5	WBBR	1130 AM	NYC	40	48	39	74	2	24	SSW	50,000	50,000	122.4	right	301	3	0	1,381	11.3
6	WHMP	1400 AM	Northampton, MA	42	19	36	72	39	26	NW	1,000	1,000	4.7	left	58	0	0	58	12.2
7	WBEC	1420 AM	Pittsfield, MA	42	26	39	73	16	43	WNW	1,000	1,000	33.0	left	20	1	0	380	11.5
																		average	11.0
																		standard deviation	0.7

Figure 16: Data sheet from a typical day's measurements – Page 1

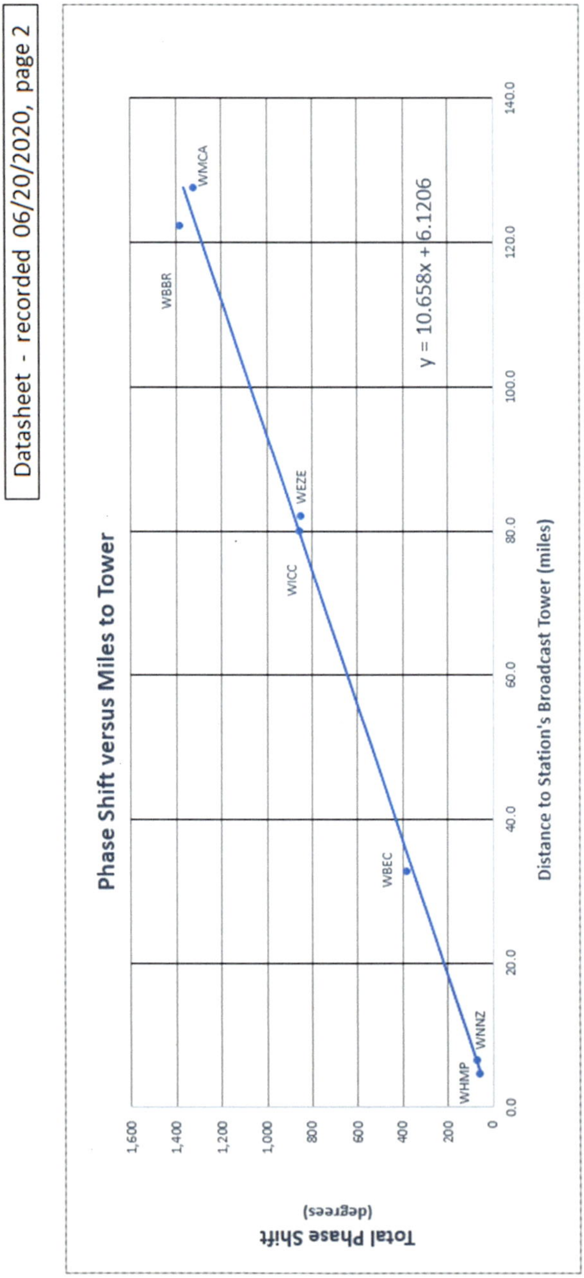

Figure 16: Data sheet from a typical day's measurements – Page 2

5 - The Math

Time of Propagation

Use the propagation of the magnetic signal as the reference and assume it travels at the speed of light in a vacuum ($C_o =$ 186,282 mi/sec), and let D represent the distance in miles it has traveled between transmitter and receiver, then the time of propagation in seconds will be

$$t = D/C_o$$

Gap Between Magnetic Signal and Electric Signal

Let d represent the distance in miles between a magnetic signal peak and the peak of the adjacent trailing electric signal, and let C_E represent the speed of propagation of the electric signal in miles per second, then

$$d = t\,(\,C_o - C_E\,)$$

$$d = (D/C_o)(\,C_o - C_E)$$

Phase Shift Between Magnetic and Electric

Let θ represent the phase shift between the magnetic and electric signals in decimal form ($0 < \theta < 1$), and let λ_M equal the wavelength of the magnetic signal in miles, then

$$\theta = d/\lambda_M$$

$$\theta = (D/C_o)(\,C_o - C_E)/\lambda_M$$

Speed of Propagation of the Electric Signal

Solving the above expression for C_E,

$$\theta = (D/C_o)(\,C_o - C_E)/\lambda_M$$

$\theta C_o \lambda_M = D(C_o - C_E)$

$(\theta C_o \lambda_M)/D = C_o - C_E$

$C_E = C_o - (\theta C_o \lambda_M)/D$

$C_E = C_o(1 - (\theta \lambda_M)/D)$ or

$C_E = C_o(1 - \lambda_M(\theta/D))$

From the experimental data we know that phase shift per mile (θ/D) is a constant (approximately) across the AM band. In degrees it was found to be 10° per mi. We can write this in decimal form as 10/360 per mi.

$$C_E = C_o(1 - \lambda_M(10/360))$$

Using this equation we can calculate C_E for the AM band and plot it. The frequency of the M and E signals are always equal.

Freq (Hz)	λ_M (mi)	C_E (mi/sec)	C_E / C_o (%)
500,000	0.373	184,354	99.0
550,000	0.339	184,529	99.1
600,000	0.310	184,675	99.1
650,000	0.287	184,799	99.2
700,000	0.266	184,905	99.3
750,000	0.248	184,997	99.3
800,000	0.233	185,077	99.4
850,000	0.219	185,148	99.4
900,000	0.207	185,211	99.4
950,000	0.196	185,267	99.5
1,000,000	0.186	185,318	99.5
1,050,000	0.177	185,364	99.5
1,100,000	0.169	185,406	99.5
1,150,000	0.162	185,444	99.6
1,200,000	0.155	185,479	99.6
1,250,000	0.149	185,511	99.6
1,300,000	0.143	185,541	99.6
1,350,000	0.138	185,568	99.6

1,400,000	0.133	185,593	99.6
1,450,000	0.128	185,617	99.6
1,500,000	0.124	185,639	99.7
1,550,000	0.120	185,660	99.7
1,600,000	0.116	185,680	99.7

Figure 17: Calculated speed of electric signal across AM band

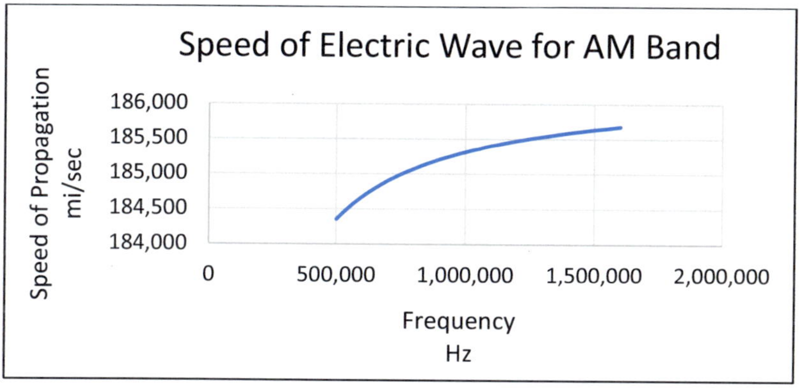

Figure 18: Calculated speed of electric signal across AM band

These are major deviations from the standard speed of light - 186,282 mi/sec. At the low end of the AM dial the formula calculates a full 1% drop.

6 - Speed of the Electric Wave

Do these lower speeds for the electric wave agree with what the literature on the subject has to say?

Two potential reasons for the slowing of electromagnetic radiation are found. Both reasons are attributed to the electric component of the waves and the way it interacts with the electric nature of the medium through which (or over which) the waves travel. They are refraction and surface waves.

Refraction

The index of refraction (n) for light traveling through air is given as 1.000293. If we want to calculate the speed of light traveling through air, we divide the speed of light by this number. Because n is so small the calculated speed of light in air is only slightly less than the speed of light in a vacuum.

> (186,282)/(1.000293) = 186,227, which is 99.97% the speed of light in a vacuum

This clearly does not explain the slower speeds of the electric signal as documented in the previous chapter.

But is it reasonable to assume that the index of refraction remains the same all the way from the visible spectrum down to the AM band? It is well known that it does not.

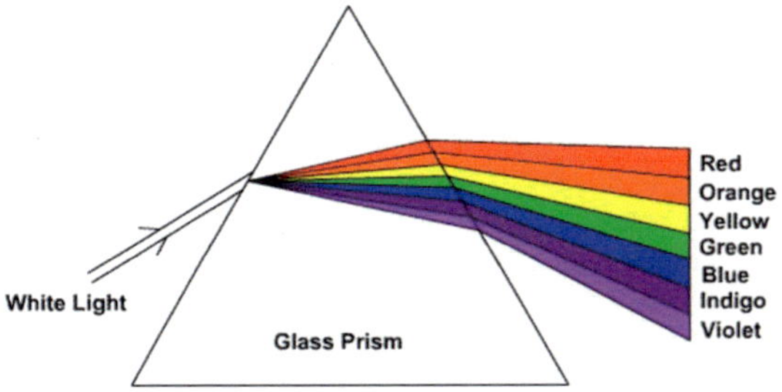

Figure 19: Dispersion of light by a prism [Ref 6]

We credit Newton with figuring out that light is a wave and when it travels through a prism the various colors that make up a beam of white light are "refracted" to a lesser or greater degree, revealing the constituent colors. Red is a lower frequency than blue and its index of refraction for the prism is slightly less than blue's so it is refracted less. The fact that the index of refraction is different at different frequencies is called dispersion.

We are told that the slower speed of light in a glass prism is due to the fact that the light interacts with electrons in the atoms of glass. The absorption and re-radiation of the light takes time resulting in an overall slower speed. The red light with its longer wavelength apparently interacts less with the atoms of glass than the blue light and travels through the glass with its speed reduced less than the blue light's speed.

So could dispersion account for the slower speeds of the electric waves? Not likely. In the visible range the dispersion that is observed is called "normal" in that the index of refraction decreases as the frequency decreases. If dispersion were causing the slowdown of the electric wave, it would be "anomalous dispersion" because the index of refraction would be increasing as the frequency decreased, not decreasing as frequency decreases as seen with the prism.

Apparently, however, dispersion exists throughout much of the electromagnetic spectrum and exhibits both types of dispersion: normal and anomalous.

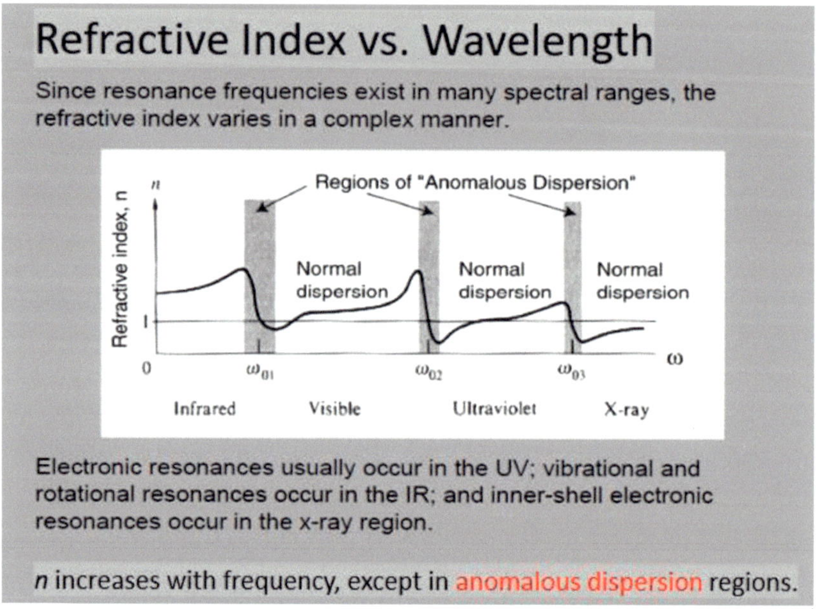

Figure 20: Refractive Index vs. Wavelength [Ref 7]

So potentially the index of refraction of air could cause the slowdown of the electric wave but n would need to be much larger for electromagnetic radiation at the AM frequency range than for visible light, and it would need to exhibit anomalous dispersion. For a 1 MHz signal the value of n needed to match the speed calculated in the prior chapter would be 1.005202, which is almost 18 times greater than the standard value of n for air.

Surface Waves

When it comes to AM broadcasting there are two types of waves. Sky waves travel more or less up and are either reflected back to earth by the ionosphere or escape into space. Ground waves travel more or less in the horizontal direction. Ground waves are said to have three components: direct waves, reflected waves and surface waves. Direct waves are "line of sight" waves. Reflected waves are waves that have reflected off mountains, large buildings or the ground. Surface waves hug the ground and in so doing have the ability to travel over the horizon. This is something that higher frequency waves do not do. The lower the frequency, the further these radio surface waves can travel while hugging the ground. AM band frequencies and below rely heavily on the surface wave phenomenon.

As the surface wave propagates in a horizontal direction it induces a flow of current in the ground, rock or water over which it passes. These currents result in a temporary buildup of charge, whose sign is opposite that of the surface wave field itself and slightly delayed because of the time required for the induced current to flow. The fields of the induced charge plus that of the original surface wave combine to create a resultant wave whose front is slightly behind the original and bent in a downward direction. [Ref 8]

This bend is what allows the wave to propagate over the horizon. Higher frequency surface waves have less time to induce a current before polarity is changed and the current must flow in the other direction and therefore do not exhibit this bending to the degree that lower frequencies do. Also, sea water is a better conductor than dry earth and consequently does a much better job of bending radio surface waves.

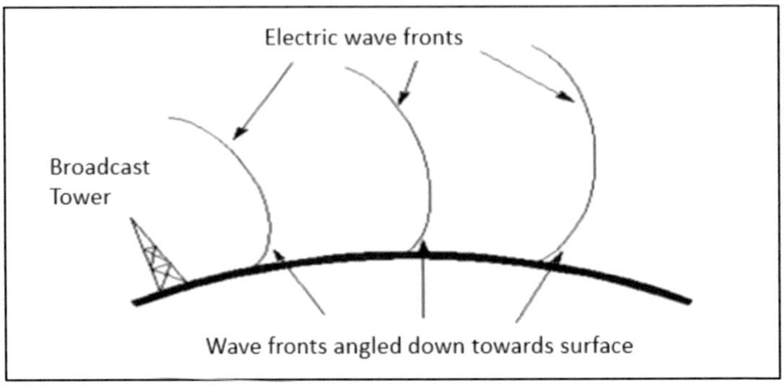

Figure 21: Radio waves bending at the surface [Ref 9]

I have found only one reference to radio surface waves that quantifies the reduction in speed. Most research is quite naturally interested in the ability of the surface wave to maintain power levels as it travels around the curvature of the earth. It is all about power not speed.

That one reference is from IEEE and dated 1950. [Ref 10] It references research done on behalf of the emerging Loran (LOng RAnge Navigation) system that was so important to military and commercial navigation after WWII up until the development of GPS in the 80's and 90's. Loran calculated location by comparing the timing of two signals received from Loran beacons, so the speed of propagation of the signals was critically important.

The reference states that a signal at 100 KHz (Loran's transmission frequency) sampled at a height of "a fraction of its wavelength" had a speed of 299,250 km/s. The wavelength of a 100 KHz signal would be 1.86 mi, or 9,821 ft. What is a "fraction" of that? 1,000 ft? That speed in miles per second is 185,945, which is 99.82 % of the speed of light.

Again this is a long way from explaining the slowdown of the electric wave calculated in the previous chapter (99.0% of the speed of light in a vacuum, at low end of AM dial). Maybe the slowdown is much greater at the surface of the earth than at

1,000 ft? But this data point is for 100 KHz, well below the AM band (550 KHz to 1600 KHz). Coming down to ground level would definitely cause a greater slowdown of the electric wave due to stronger induced currents, but moving to the higher AM frequencies would work against this being the explanation we are looking for.

The radio surface wave explanation is intriguing because the surface wave phenomenon is so closely tied to frequency. Cut the frequency in half and of course the wavelength doubles. A doubled wavelength doubles the time that the surface wave gets to induce a positive or negative charge in the surface. That means a stronger induced wave and a slower resultant wave.

Surface Waves

These two diagrams illustrate the following:

1. The induced electric wave has a charge opposite to that of the surface wave
2. The induced electric wave propagates slightly behind the surface wave
3. The longer the wavelength the deeper the induced wave reaches into the surface
4. The deeper the induced wave the greater the buildup of charge
5. The greater the buildup of charge the slower the resultant wave near the surface

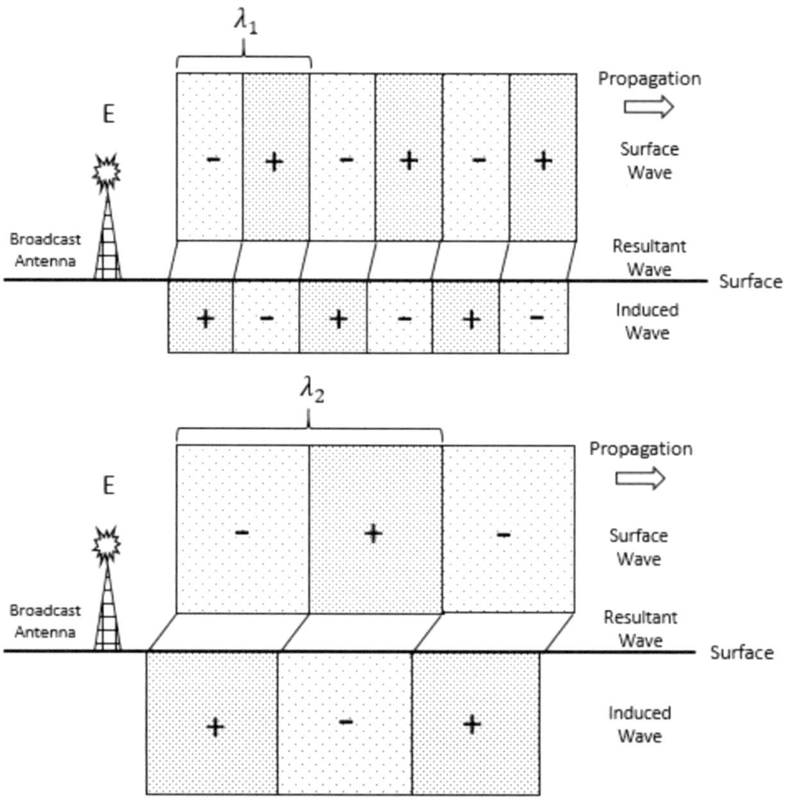

Figure 22: Surface Waves

This could explain the tight correlation between longer wavelength and slower electric wave speed that results in a constant phase shift per mile across the full AM band. It would be interesting to test the electric to magnetic phase shift at different elevations (think tethered hot air balloon) and across water (think Long Island Sound).

Another factor that makes the surface wave explanation attractive is the fact that the speed of the electric wave does not seem to depend on weather. I included a place to record weather on my data sheets thinking that some day I might see a correlation between the weather and the phase shift. It seems reasonable that temperature, humidity, rain, snow and barometric pressure might affect the speed of the electric, but the data shows they do not. (Keep in mind my tolerance on any given phase shift measurement is up to 5°.)

This makes sense from a surface wave perspective. The passing electric wave induces a current flow and subsequent charge in the ground over which it passes. These changes are taking place down to hundreds of feet below the surface. At really low frequencies the surface wave phenomenon makes it possible to have radio communication with deep mines and submarines. So it is the conductivity of the earth well below the surface that gives us surface waves, not the weather.

Nevertheless

Even though the mechanism behind the reduction of the electric wave's speed has not been fully identified, the major finding of these experiments still stands. The speeds of propagation for the electric and magnetic components of a given AM signal differ and this is enough to establish the fact that they travel separately and do not interact once they have left the broadcast tower.

7 - Full Accounting for Phase Shift

The phase shift discussed so far has been that which takes place while the electromagnetic wave is propagating from transmitter to receiver. It is helpful to keep in mind that phase shift also occurs before and after propagation.

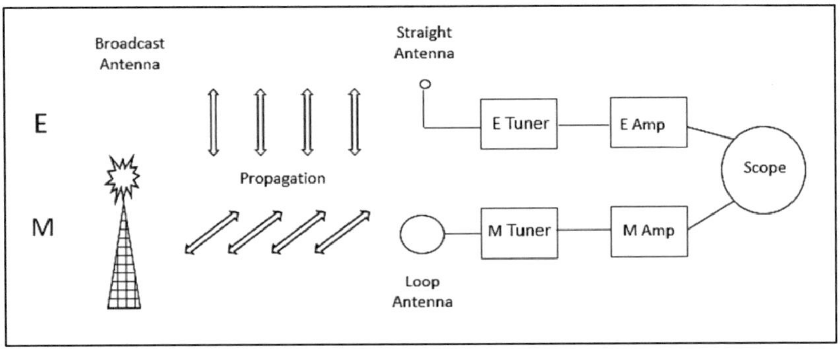

Figure 23: The different sources of phase shift from transmitter to oscilloscope

Broadcast Antenna

At the broadcast antenna, the antenna starts at a neutral electric potential, builds to a peak positive electric potential and returns to neutral. Then it builds to a peak negative potential and returns to neutral again. The magnetic field that encircles the antenna with north pointing to the left, for example, will peak when the current flow in the antenna is at its peak, not when the potential (voltage) is at its peak. At peak electric potential the current has actually stopped momentarily as it changes direction, flowing back the other way and reaching its peak flow as it passes through the zero electric potential point.

This means that the current peak will follow the voltage peak by 90 degrees. So when the electromagnetic wave leaves the broadcast tower the magnetic component is 90 degrees behind the electric component.

Propagation

The E drops behind the M at a rate of 10 degrees per mile.

Receiving Antenna - E

When a positive electric field reaches the E receiving antenna, electrons are drawn up into it amounting to a 180 degree phase shift. For a negative electric field the electrons are chased down the antenna. So at the receiving antenna the E field goes from being a propagating E field to an electrical voltage in a wire and picks up a 180 degree phase shift.

Receiving Antenna - M

When a leftward swinging M field (for example) passes through the loop antenna, it will be the changing M field that induces a voltage in the antenna. At peak M, the change in the M field is actually zero as it swings from a leftward direction to a rightward direction. It is 90 degrees later when the M field is experiencing its greatest rate of change as it passes through the zero M field point. So at the antenna, the M field goes from being a propagating M field to an electrical voltage in a wire and picks up a 90 degree delay.

Tuning

Rough and fine tuning must be done very carefully. You can actually see the phase shift walk forward or backwards on the scope with just a slight twist of the fine tuning knob, especially at the high end of the AM band. A perfectly tuned tuner will introduce zero phase shift. If you cannot tune both radios to a steady peak then you cannot measure the phase shift for that signal.

Amplification

Some transistor circuits will add 180 degrees of phase shift to the signal and others will not. There is always some stray capacitance in the circuit that will shift the signal. Both of these can be overcome by using identical designs for the two radios. As mentioned earlier, the tuning coils are used as the antennas in both radios - the M coil in a horizontal position and the E coil in a vertical position.

To verify that the two amplifiers are not adding different amounts of phase shift, put both coils in the same orientation (both horizontal or both vertical) and tune to the same station. At this point the phase shift should measure zero. (My standard is less than 5 degrees.) This needs to be checked at both the high and low ends of the AM band.

Oscilloscope

The two signals are displayed on a two channel oscilloscope where the final phase shift is measured. Any phase shift introduced by the scope itself will be the same for both channels and will cancel out. Defective leads on the scope can be a problem but can be detected by swapping.

Summarizing the above

θ	Broadcast	Propagate	Receive	Tune	Amplify	Scope	Total
M	-90	0	-90	0	0	0	-180
E	0	(- 10°/mi)	-180	0	0	0	-180

Figure 24: Summary of Phase Shift Sources

With the amplifiers balanced and the tuners carefully tuned, any phase shift measured on the scope can be attributed to the different propagation speeds of the M and E waves.

8 - Oleg Dmitrovitch Jefimenko

During the last ten years of my father's life, the usage of the internet exploded with what seemed like a magical ability to bring us answers to any question on our minds. From time to time I would do a search to see if there was anything out there relating to independent electric and magnetic field propagation. These were not sophisticated searches on my part because I had no idea where to look. The answer was always no. One year after my father passed away I found a reference to Oleg Dmitrovitch Jefimenko.

Jefimenko (1922-2009) was a physicist and a professor. He received his Ph.D. from the University of Oregon and taught at West Virginia University. With a deep understanding of the math and physics involved in electromagnetic radiation, he proposed that the "causal" relationship between the electric and magnetic components of electromagnetic waves did not exist. And further, he claimed that Maxwell never actually said it did. He said they share a common origin but that is all.

> *"1-5. Can Time-Variable Electric and Magnetic Fields Create Each Other?*
>
> *There is a widespread belief that time-variable electric and magnetic fields can cause each other. The analysis of Maxwell's equations presented above does not support this belief. It is true that whenever there exists a time-variable electric field, there also exists a time-variable magnetic field. This follows from Maxwell's Eqs. (...). But, as we have seen, neither Maxwell's equations nor their solutions indicate an existence of causal links between*

> *electric and magnetic fields. Therefore we must conclude that an electromagnetic field is a dual entity always having an electric and a magnetic component simultaneously created by their common sources: time-variable electric charges and currents."* [Ref 11]

I am not qualified to say whether Jefimenko's analysis is correct but it does fit the data that we have seen. I wish that my dad could have read Jefimenko's words.

9 - Additional Thoughts

Bouncing, Skipping and Reflecting

Could this phase shift phenomena be caused by the AM radio waves bouncing, skipping or reflecting? Yes, conceptually this could explain it.

A given radio's signal could take two different paths in order to arrive at two receivers sitting side by side. One path might be straight line of sight and the other involve a reflection off the side of a large building. The second path would be longer than the first so the signal that took that path would arrive a short time after the first. With one radio receiving the first signal and the other radio receiving the second signal, the two signals would appear phase shifted when displayed on the oscilloscope one above the other.

But these multipath scenarios are rare. The transmitter, the building and the two receivers would have to line up just right at the right distances and with the right frequencies to make it happen. It would be virtually impossible to find a location where the signals from all the stations across the AM dial would arrive via multiple paths and produce a phase shift of 10 degrees per mile from their respective transmitters, as seen in the data.

Circular Polarization

Another possible explanation that comes to mind is circular polarization. In AM broadcasting, the electric field is oriented vertically while the magnetic is oriented horizontally. It is assumed that these orientations are retained throughout propagation. Circular polarization is a phenomenon that has been observed in certain materials and in outer space that causes the orientations to rotate while propagating. So that after a given distance of travel, the electric field is no longer oriented vertically, for example, but has rotated a number of degrees clockwise or counterclockwise with the magnetic field rotating with it.

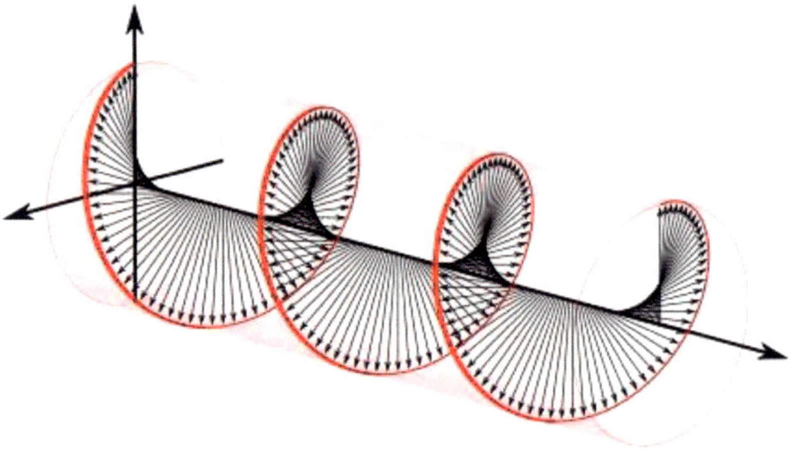

Figure 25: Circular Polarization [Ref 12]

If the AM signal were undergoing circular polarization during propagation there would be no impact on the signal being picked up by the electric radio. This is because the electric antenna is just a wire completely immersed in the positive or negative field of the signal without regard to the “direction” of the electric field. (The wavelength of a 1 MHz AM signal is on the order of .186 miles or 984 feet long.) The magnetic signal, however, would be another matter.

Assuming the antenna coil on the magnetic radio is lying horizontally and turned sideways to the transmitter so the maximum amount of magnetic field is sweeping through the coil’s windings, any amount of rotation of the horizontal magnetic field caused by circular polarization during propagation would decrease the strength of the signal received. Potentially this decrease in strength could be mistakenly taken for a phase shift between the electric and the magnetic.

I tested for circular polarization and I have not seen it. With the coil in the horizontal plane, aligned sideways to the transmitter for maximum signal, I have lifted one end of the coil and then the other end. This effectively rotates the coil in the vertical plane that faces the transmitter in clockwise and counterclockwise directions. If circular polarization was taking place, the signal strength would increase with one or the other of these moves because the alignment of the open center of the coil to the magnetic field would improve. I have done this on a number of occasions and in each case the signal decreased. Therefore, no circular polarization.

Why Overlooked for So Long?

Basically, the separate nature of the electric and magnetic components of electromagnetic waves has been hiding in plain sight for the past 150 years. Why has it gone unobserved for so long? Perhaps it is because detecting the electrical component is so much easier than detecting the magnetic. The electric field will interact with any charged particle or the valence electrons of any atoms. Electrons are everywhere. And the force of an electric field on an electron is very strong.

Nature has taken advantage of this fact in the evolution of photosynthesis and eyesight. To detect the magnetic field, on the other hand, would require a flowing current and the detection of the deflection of that current to one side or the other caused by the magnetic field. It is not surprising that nature found an electric field hitting an atom's outer electrons and triggering a chemical reaction much simpler and more powerful than working with the magnetic field. Likewise it is not surprising that humans have developed chemical photography, radio and cell phone communication based upon the electric field.

Detecting the magnetic field by current flow deflection is known to us as the Hall effect. It typically involves the detection of nearly static or low frequency magnetic fields and has a comparatively small number of uses. The detection of the magnetic field at visual frequencies is just now being developed and is not generally available.

We know that birds, dogs and whales can navigate using the earth's magnetic field so clearly they have developed the biological mechanisms to detect magnetic fields. But this is detection of a virtually static magnetic field. Nowhere do we see detection of the magnetic side of electromagnetic radiation.

In general, the use of magnetic fields in nature and in human technology is far secondary to the use of electric fields. If the job can be done better and easier with the electric field why waste time developing magnetic field detection? And if historically the focus has been exclusively on the electric side, maybe that could explain why the slightly different speeds of the two components has gone undetected.

Visual range

If the magnetic and electric fields of an AM band transmission travel independently, it is reasonable to think that the same happens at visual frequencies. If so, we have two types of light - one that we are very familiar with and another that we have never seen. When we think of the behavior of light we think of reflection, refraction, dispersion, diffraction and transparency. We think of color. All of these characteristics can be traced back to the interaction of the electric field of light with the electric field of the valence electrons of the atoms that make up the medium. Light as we know it is an electric effect. The magnetic component of light is nowhere to be found.

So how do reflection, refraction, dispersion, diffraction and transparency behave when the wave is magnetic? This is a fascinating question. My dad imagined that someday much larger earth based telescopes could be built if they utilized the magnetic field because magnetic starlight coming through the atmosphere would not experience scintillation as the electric startlight does. Presumably magnetic starlight would not "twinkle".

10 - Conclusion

I have redone my dad's experiment Number 7 and have come to the same conclusion. The electric and magnetic components of an electromagnetic radio wave are NOT joined at the hip. They do not rely on one another for their propagation. They are generated by a common source but once launched they travel independently. And because, as it turns out in the case of AM surface waves, the electric component travels at a slightly slower speed than the magnetic component, there is a measurable phase shift between the two waves when they arrive at the receiver.

The magnetic component seems to maintain its speed at the speed of light, but this is more of an assumption than a finding. The electric component travels at a lower speed than the magnetic but that speed is not constant - it drops as the frequency drops.

Surprisingly, the rate of drop in the electric component's speed is just right so that when the difference in the arrival time of the magnetic and electric components is expressed as a phase shift, a constant phase shift per mile emerges regardless of the station's broadcast frequency or propagation distance. That rate is approximately 10°/mi.

Using that rate we can derive an expression for the speed of the electric component:

$$C_E = C_o(1 - \lambda_M(10/360))$$

The receivers and their antennas are not the cause of the observed phase shift. They could not be because they are blind as to how far the signals have traveled. All the receivers can tell us is that the magnetic and electric signals of a given AM station arrive at different times. But from that difference we can calculate how far those signals have traveled and thus how far away the transmitter is. This could only be true if the electric and magnetic signals propagated independently.

Appendix 1: Schematic for Dad's Radios

Here is a schematic of my dad's radio for the magnetic signal. The electric radio had the same construction except for the antenna which was a straight wire not a loop.

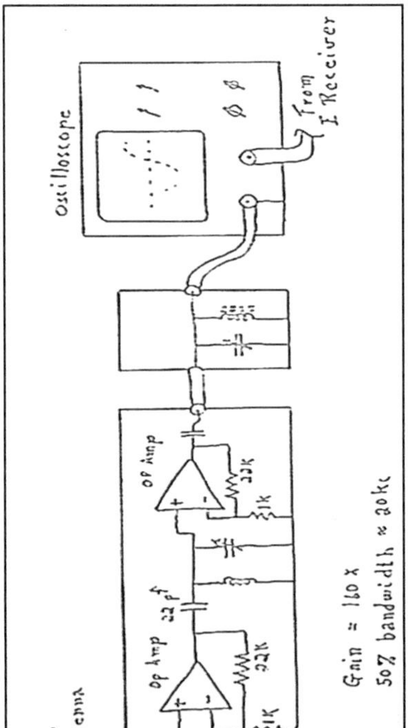

Figure 26 Schematic for Dad's radios

I am not sure which op amps he used. They had to be able to handle the frequency at the top of the AM band which is 1.6 MHz. The op amps I used in my radios were AD847.

In this schematic you can see that he wrapped aluminum foil around the loop antenna and then grounded the foil. This makes sense because although the loop is designed to pick up the magnetic signal and does a good job of that, there is nothing to keep it from also picking up the electric. Thus the grounded foil intercepts the electric while allowing the magnetic to pass only slightly attenuated. The electric antenna, on the other hand, will only pick up the electric signal.

Conceptually, having two signals from the same station arriving on the loop antenna out of phase (one electric and one magnetic) would lead to confusing results in the phase measurement. I worried about this and made numerous attempts to shield my loop from the electric but never found any real benefit. In the end I believe the fact that the magnetic signal was typically much stronger than the electric saved me. I am guessing that although both electric and magnetic use the same coils in their tuning tanks, the manner in which the magnetic field induces current flow in the coil somehow gives it an advantage over the electric.

Dad used sheet iron to insulate the radios from outside interference. I used standard aluminum project boxes. I struggled with interference and wonder if I should have used the iron. The power adaptor for my laptop computer and the laptop computer itself were a real problem with AM frequency interference until I identified the source.

Appendix 2: Schematic for My Radios

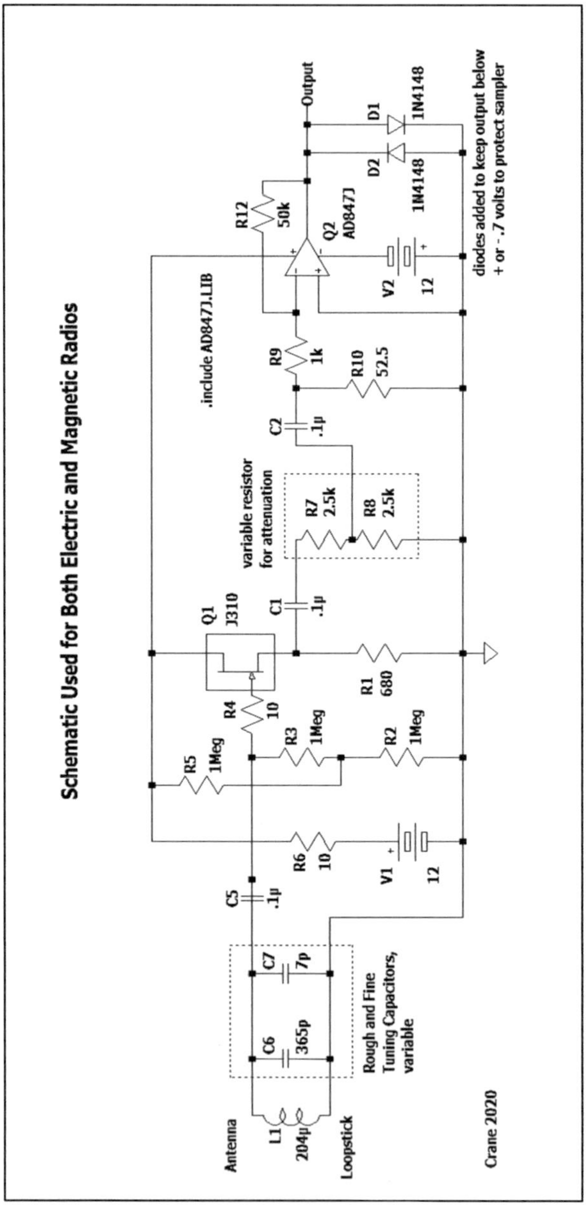

Figure 27: Schematic for my radios

The JFET transistor (Q1) was chosen because of its wonderfully high input impedance that did not draw down the signal voltage coming off the antenna.

The op amp was added for gain but was not really need it.

The two diodes at the output basically do nothing until the signal goes above + or - 0.7 Volts, at which point they short to ground and limit the signal from getting any stronger. They were added after frying a couple of preamps. The Digilent A-to-D sampler fed by the receivers claims to be able to handle 25 Volts but I wanted to be sure.

These radios have some inherent phase shift between them but because their design is identical most of that is effectively cancelled. I verify this fact by doing what I call a Magnetic-to-Magnetic (M-to-M) calibration. An M-to-M calibration is done by placing the tuning-antenna coils on both radios in a horizontal orientation and tuning both radios to the same station. With the radios carefully tuned there should be zero phase shift between them because of course they are both tuned to the same magnetic signal. My goal is to have the inherent phase shift between the radios be less than 5 degrees. If there is more than 5 degrees I make a note of it on the data sheet in the Initial Offset column and that difference is automatically subtracted from the measured phase shift. An E-to-E calibration is just as valid.

Appendix 3: Loopstick Design

I spent a lot of time experimenting with loop and loopstick antennas. My goal was to come up with an antenna for the magnetic signal that would maximize the following.

1. Pick up a strong signal
2. Have an inductance that when paired with the standard 365 pico Farad tuning coil would allow tuning across the full AM band of 550 to 1600 KHz
3. Maximize the selectivity
4. Minimize inherent phase shift between M and E radios by using one antenna design on both radios if possible

Here are pictures of the loops and loopsticks tested for best design.

Figure 28: Loop antennas evaluated

Figure 29: Loopstick antennas evaluated

The different parameters of design included:

1. Wire type – solid copper, flat copper foil, litz wire
2. Wire size (diameter)
3. Loop diameter
4. Number of turns
5. Spacing between turns
6. Number of ferrite rods
7. Length of ferrite rod
8. Composition of ferrite rod

The magnetic antenna chosen was a loopstick that had a core of seven ferrite rods (each rod 5 ½ by 5/16 inches) wrapped with 50 turns of litz wire. The ferrite rods had a permeability rating of 400. The litz wire had a rating of 1162/46 meaning it consisted of 1,162 fine copper thread-like conductors insulated from one another by enamel, each with an AWG rating of 46. AWG 46 is a very thin conductor, .0016 inches in diameter.

The goal of litz wire is to reduce resistance by providing a wire with effectively a greater cross sectional area for the electrons to pass through. Because of the repulsion between electrons, a current flowing in a solid copper wire will restrict its travel to the outer edges of the wire – this way each electron will be as far away as possible from all other electrons. Not sure why the electrons in the litz wire do not notice the current in neighboring threads and refuse to travel in all but the outer threads but it seems to work as advertised, cutting the resistance in half which is very helpful for the Q. The measured inductance of the loopstick was 204 micro Henrys.

I am not convinced that the extra ferrite rods in the core and the use of litz wire was worth the effort.

Appendix 4: Wraparound

As previously mentioned, the name my dad gave to the fact that a given station's total phase shift was more than 360° was "wraparound". For example, if a station was 50 miles away, the measured phase shift would be 140°. But the total phase shift was actually 140° plus 360°, or 500°. So the phase shift per mile would be 500°/50mi, which is the expected 10°/mi.

Adding the extra wraparound for stations beyond 36 miles makes sense, but only up to a point. The problem is that when you get to stations that are 3 or 4 wraparounds away the total phase shift consists primarily of the wraparound phase shift calculated from the known distance to the transmitter, with the measured phase shift making up a smaller and smaller portion of the total. Consequently, the stations at 100 miles distace or more add no real value to the investigation of the overall average phase shift per mile. Their data point is basically a fabricated data point, not a true measured data point.

I have included stations at 3 and 4 wraparounds in my datasheets because these stations are out there and I can measure their phase shift. But keep in mind that as they get further and further away they are more or less along for the ride and do not really contribute to a better understanding of the overall average phase shift per mile number.

In the datasheet on the next page I have entered some ficticious data points to illustate this.

Ficticious Data Wraparound Demo

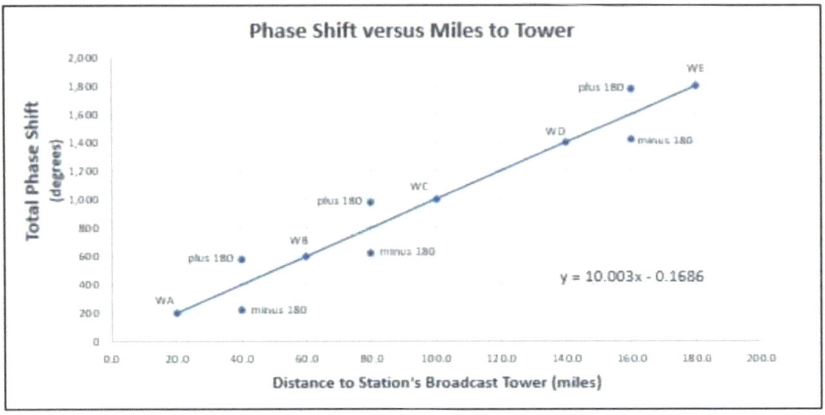

	Station	Freq (KHz)	City	Distance (mi)	Measured Phase Shift (degrees)	Wrap-arounds	Initial Offset (degree)	Total Phase Shift (degree)	Phase Shift / Mile (degree)	Quality of Reading
1	WA			20.0	200	0	0	200	10.0	
	plus 180			40.0	40	1	-180	580	14.5	+4.5
	minus 180			40.0	40	1	180	220	5.5	-4.5
2	WB			60.0	240	1	0	600	10.0	
	plus 180			80.0	80	2	-180	980	12.3	+2.3
	minus 180			80.0	80	2	180	620	7.8	-2.2
3	WC			100.0	280	2	0	1,000	10.0	
4	WD			140.0	320	3	0	1,400	10.0	
	plus 180			160.0	160	4	-180	1,780	11.1	+1.1
	minus 180			160.0	160	4	180	1,420	8.9	-1.1
5	WE			180.0	1	5	0	1,801	10.0	
									10.0	average phase shift
									2.3	standard deviation

The stations numbered 1 through 5 represent five ideal stations showing exactly 10°/mi phase shift at 20, 60, 100, 140 and 180 miles. The three pairs of shaded rows represent measurements made with + and - 180° of measurement error at 40, 80 and 160 mi. (I am using the "Initial Offset" column to add the errors.)

Notice how the first pair of shaded rows is at 1 wraparound and, with 180° of measurement error, gives readings that are + and - 4.5 from the average. (A deviation that clearly shows that these data points are not to be trusted.) The third pair of shaded rows at 4 wraparounds, also with 180° of measurement error, gives readings at + and -1.1 from the averge. (A deviation that makes these data points look valid when infact they are 180° off.)

This demonstrates that measurements with a high number of wraparounds are of little value because their total phase shift is primarily based upon the known distance to the transmitter.

Appendix 5: Data

Datasheet - recorded 07/08/2020, page 1					
Northampton, MA					
Latitude			Longitude		
42	15	39.1	72	40.6	60.0
42° 15' 39.1'' N			72° 40.6' 60'' W		

Start Time	5:00 PM	AM/PM
Stop Time	7:00 PM	AM/PM
Temp	83	F
Humidity	63	%
Ground	dry	dry/wet/snow
Air	dry	dry/rain/snow
Pressure	29.92	inHg
Wrap-around	29.2	mi

	Station	Freq (KHz)	City	Lat			Long			Direction	Power Day (Watts)	Power Night (Watts)	Distance (mi)	Red Wire on M Coil, outbound	Measured Phase Shift (degrees)	Wrap-arounds	Initial Offset (degree)	Total Phase Shift (degree)	Phase Shift / Mile (degree)
1	WNNZ	640 AM	Westfield, MA	42	10	46	72	45	3	SW	5,000	1,000	6.4	right	190	0	90	100	15.7
2	WSBS	860 AN	Greatbarrington MA	42	12	52	73	20	45	W	2,700	4	33.6	left	344	0	0	344	10.2
3	WHMP	1400 AM	Northampton, MA	42	19	36	72	39	26	NW	1,000	1,000	4.9	left	54	0	0	54	11.0
																		average	12.3
																		standard deviation	3.0

Datasheet - recorded 07/08/2020, page 2

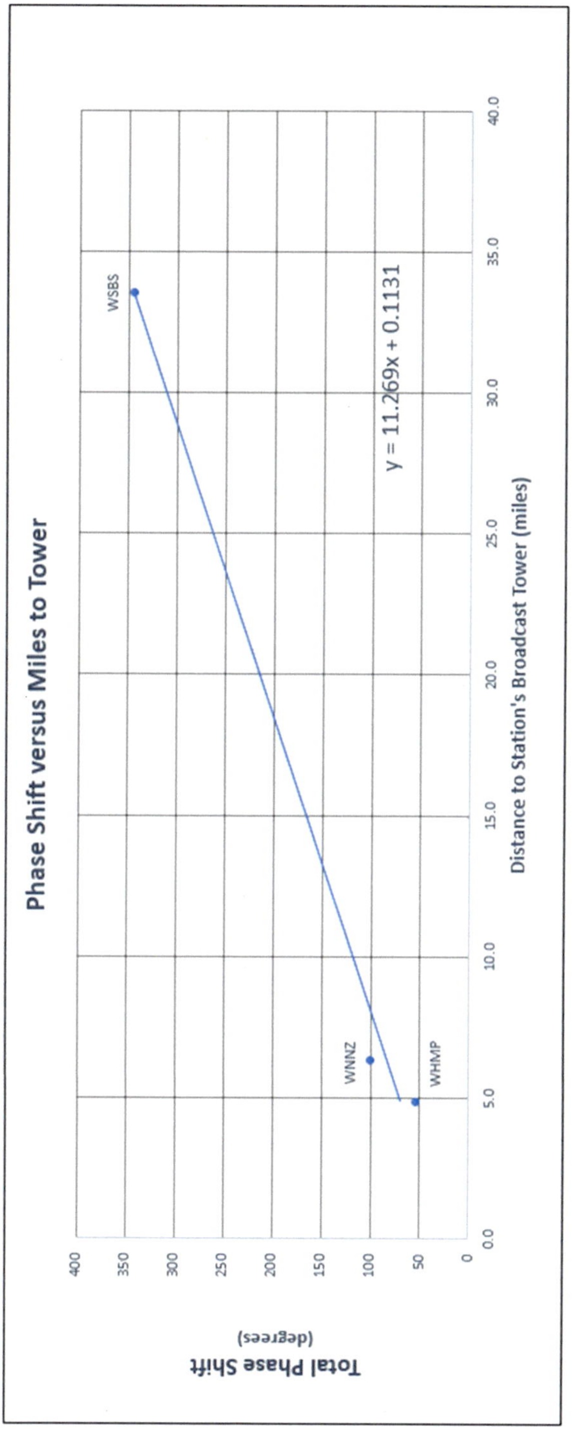

Start Time	6:45 AM	
Stop Time	8:30 AM	
Temp	66	F
Humidity	90	%
Ground	dry	dry/wet/snow
Air	dry	dry/rain/snow
Pressure	30.05	inHg
Wrap-around	32.8	mi

Datasheet - recorded 06/20/2020, page 1					
Easthampton, MA					
Latitude			Longitude		
42	15	39.1	72	40	59.8
42° 15' 39.1'' N			72° 40' 59.8'' W		

	Station	Freq (KHz)	City	Lat			Long			Direction	Power Day (Watts)	Power Night (Watts)	Distance (mi)	Red Wire on M Coil, outbound	Measured Phase Shift (degrees)	Wrap-arounds	Initial Offset (degree)	Total Phase Shift (degree)	Phase Shift / Mile (degree)
1	WMCA	570 AM	NYC	40	45	10	74	6	15	SSW	5,000	5,000	127.7	left	240	3	0	1,320	10.3
2	WEZE	590 AM	Boston, MA	42	24	24	71	5	14	E	5,000	5,000	82.2	right	128	2	0	848	10.3
3	WICC	600 AM	Bridgeport CT	41	9	36	73	9	43	SSW	1,000	500	80.1	right	134	2	0	854	10.7
4	WNNZ	640 AM	Westfield, MA	42	10	44	72	45	4	SSW	50,000	1,000	6.6	right	160	0	90	70	10.5
5	WBBR	1130 AM	NYC	40	48	39	74	2	24	SSW	50,000	50,000	122.4	right	301	3	0	1,381	11.3
6	WHMP	1400 AM	Northampton, MA	42	19	36	72	39	26	NW	1,000	1,000	4.7	left	58	0	0	58	12.2
7	WBEC	1420 AM	Pittsfield, MA	42	26	39	73	16	43	WNW	1,000	1,000	33.0	left	20	1	0	380	11.5
																		average	11.0
																		standard deviation	0.7

Datasheet - recorded 06/20/2020, page 2

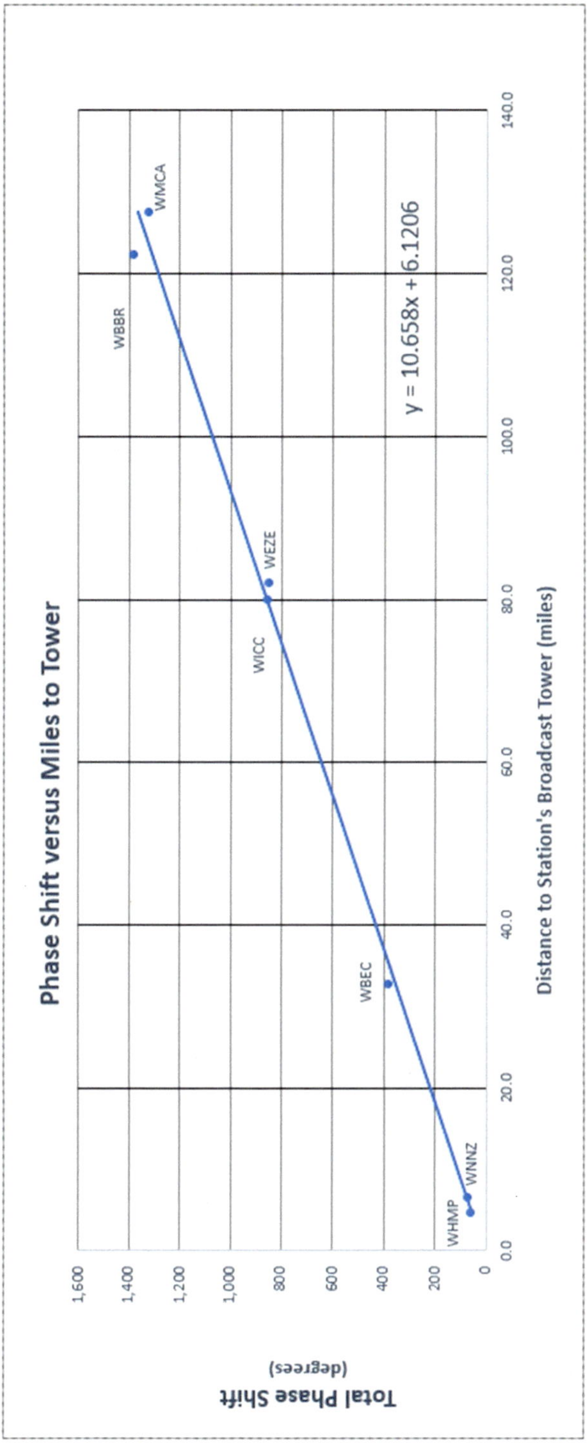

Start Time	10:30 AM	AM/PM
Stop Time	12:00 PM	AM/PM
Temp	48	F
Humidity	39	%
Ground	dry	dry/wet/snow
Air	dry	dry/rain/snow
Pressure	30.00	inHg
Wrap-around	37.5	mi

Datasheet - recorded 05/10/2020, page 1					
Easthampton, MA					
Latitude			Longitude		
42	15	39.1	72	40	59.8
42° 15' 39.1'' N			72° 40' 59.8'' W		

	Station	Freq (KHz)	City	Lat			Long			Direction	Power Day (Watts)	Power Night (Watts)	Distance (mi)	Red Wire on M Coil, outbound	Measured Phase Shift (degrees)	Wrap-arounds	Initial Offset (degree)	Total Phase Shift (degree)	Phase Shift / Mile (degree)
1	WHYN	560 AM	Springfield (Westfield)	42	11	37	72	41	0	S	5,000	1,000	4.6	right	45	0	0	45	9.7
2	WMCA	570 AM	NYC	40	45	10	74	6	15	SSW	5,000	5,000	127.7	right	109	3	0	1,189	9.3
3	WEZE	590 AM	Boston, MA	42	24	24	71	5	14	E	5,000	5,000	82.2	right	20	2	0	740	9.0
4	WICC	600 AM	Bridgeport CT	41	9	36	73	9	43	SSW	1,000	500	80.1	right	102	2	0	822	10.3
5	WNNZ	640 AM	Westfield, MA	42	10	44	72	45	4	SSW	50,000	1,000	6.6	right	155	0	90	65	9.8
																		average	9.6
																		standard deviation	0.5

Datasheet - recorded 05/10/2020, page 2

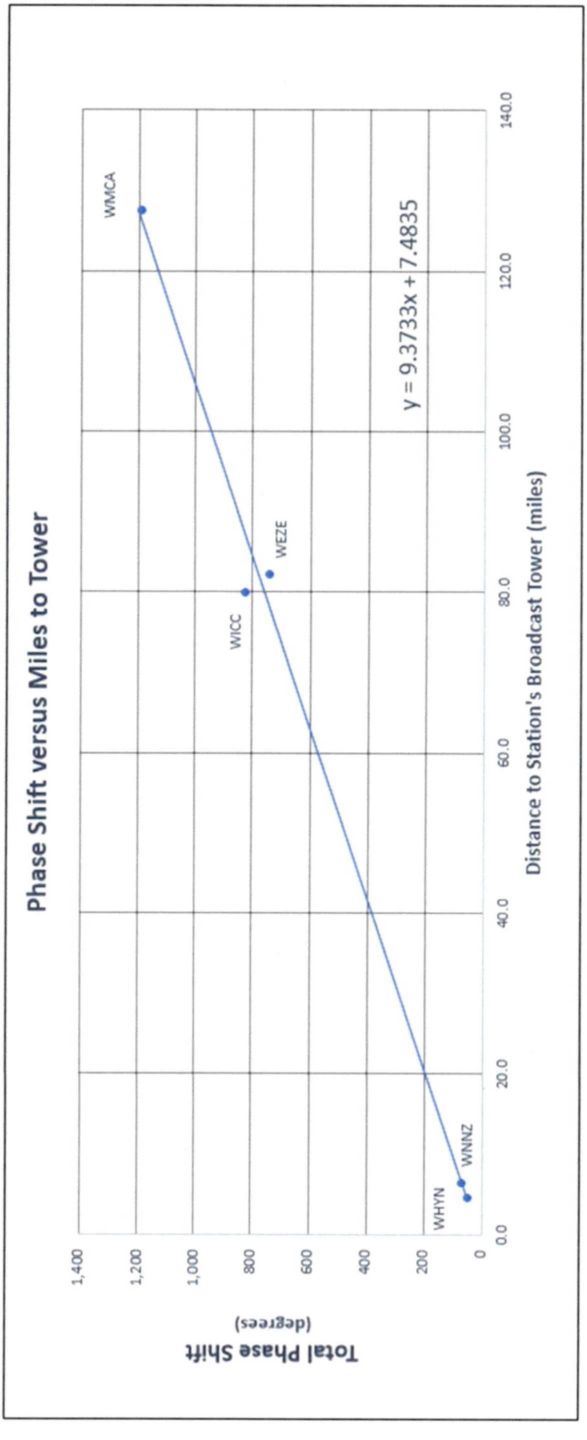

Start Time	5:45 AM	AM/PM
Stop Time	7:30 AM	AM/PM
Temp	58	F
Humidity	38	%
Ground	dry	dry/wet/snow
Air	dry	dry/rain/snow
Pressure	~	inHg
Wrap-around	35.0	mi

Datasheet - recorded 05/04/2020, page 1					
Easthampton, MA					
Latitude			Longitude		
42	15	39.1	72	40	59.8
42° 15' 39.1'' N 72° 40' 59.8'' W					

	Station	Freq (KHz)	City	Lat			Long			Direction	Power Day (Watts)	Power Night (Watts)	Distance (mi)	Red Wire on M Coil, outbound	Measured Phase Shift (degrees)	Wrap-arounds	Initial Offset (degree)	Total Phase Shift (degree)	Phase Shift / Mile (degree)
1	WEZE	590 AM	Boston, MA	42	24	24	71	5	14	E	5,000	5,000	82.2	right	188	2	0	908	11.0
2	WACE	730 AM	Chicopee, MA	42	10	2	72	37	29	SSE	5,000	8	7.1	right	72	0	0	72	10.1
3	WSBS	860 AM	Great Barrington, MA	42	12	52	73	20	45	W	2,700	4	34.1	left	320	0	0	320	9.4
4	WBBR	1130 AM	NYC	40	48	39	74	2	24	SSW	50,000	50,000	122.4	right	168	3	0	1,248	10.2
5	WHMP	1400 AM	Northampton, MA	42	19	36	72	39	26	N	1,000	1,000	4.7	left	51	0	0	51	10.8
																		average	10.3
																		standard deviation	0.6

Datasheet - recorded 05/04/2020, page 2

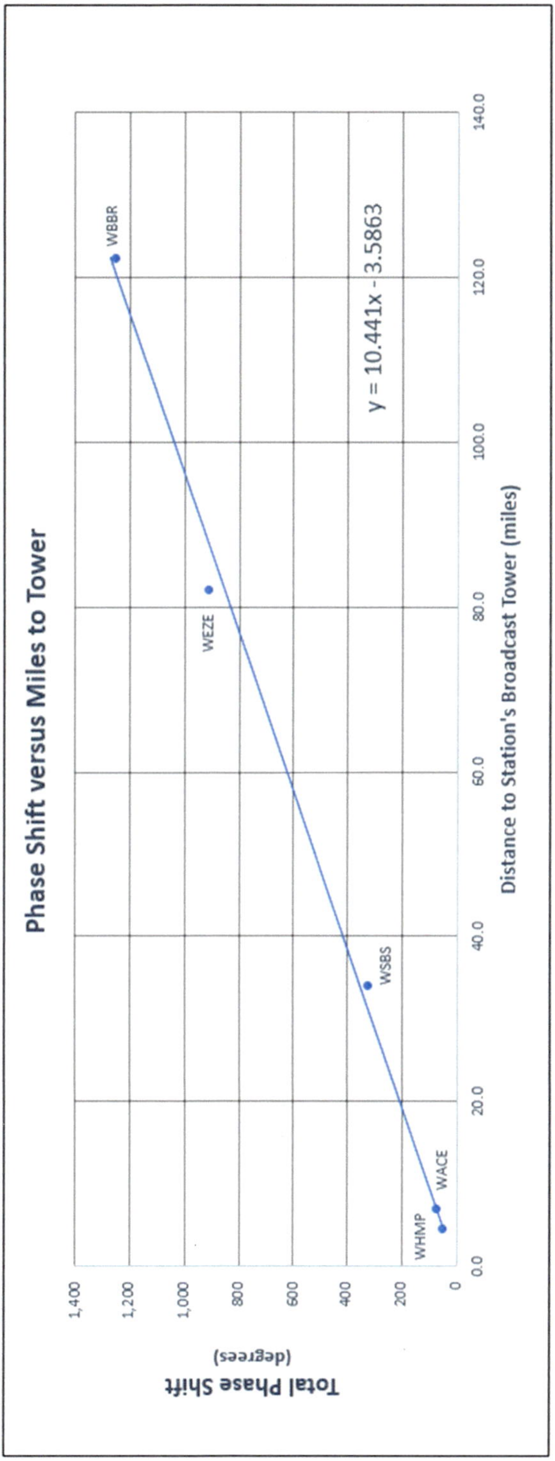

Start Time	6::00 AM	AM/PM
Stop Time	7:45 AM	AM/PM
Temp	53	F
Humidity	73	%
Ground	dry	dry/wet/snow
Air	dry	dry/rain/snow
Pressure	29.79	inHg
Wrap-around	34.5	mi

Datasheet - recorded 5/3/2020, page 1					
Easthampton, MA					
Latitude			Longitude		
42	15	39.1	72	40	59.8
42° 15' 39.1" N			72° 40' 59.8'' W		

	Station	Freq (KHz)	City	Lat			Long			Direction	Power Day (Watts)	Power Night (Watts)	Distance (mi)	Red Wire on M Coil, outbound	Measured Phase Shift (degrees)	Wrap-arounds	Initial Offset (degree)	Total Phase Shift (degree)	Phase Shift / Mile (degree)
1	WEZE	590 AM	Boston, MA	42	24	24	71	5	14	E	5,000	5,000	82.2	right	217	2	0	937	11.4
2	WACE	730 AM	Chicopee, MA	42	10	2	72	37	29	SSE	5,000	8	7.1	right	68	0	0	68	9.5
3	WSBS	860 AM	Great Barrington, MA	42	12	52	73	20	45	W	2,700	4	34.1	left	350	0	0	350	10.3
4	WBBR	1130 AM	NYC	40	48	39	74	2	24	SSW	50,000	50,000	122.4	right	175	3	0	1,255	10.2
5	WHMP	1400 AM	Northampton, MA	42	19	36	72	39	26	N	1,000	1,000	4.7	left	51	0	0	51	10.8
																		average	10.4
																		standard deviation	0.7

Datasheet - recorded 5/3/2020, page 2

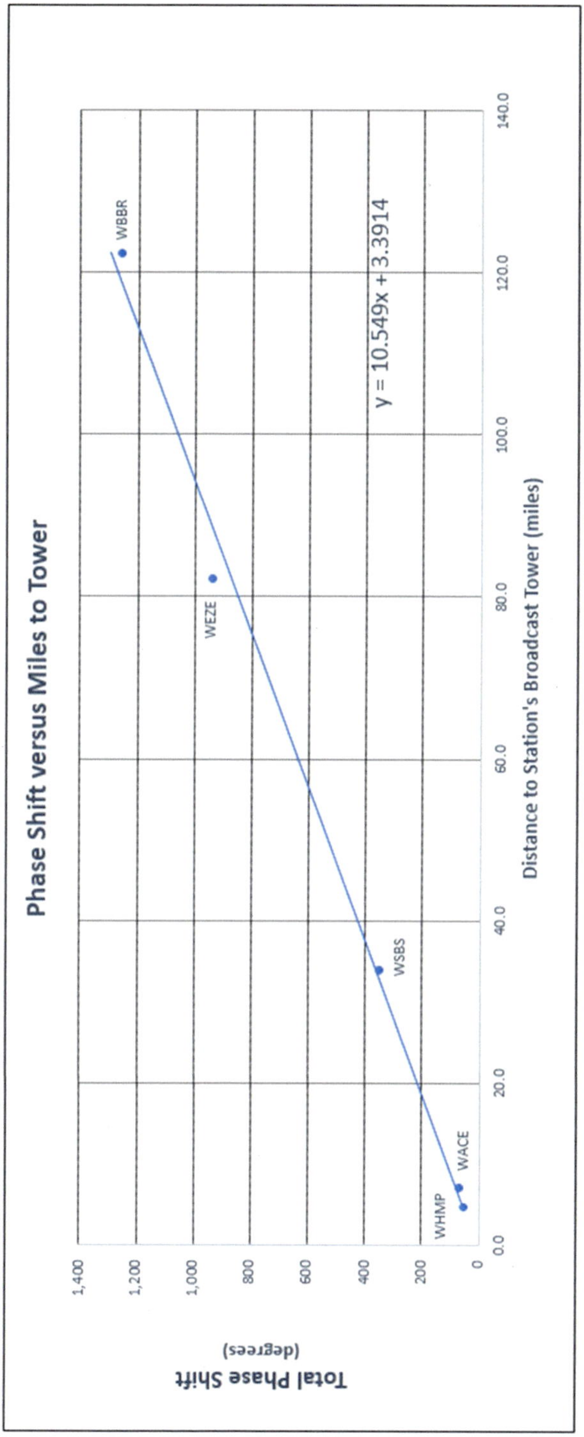

Start Time	10:20 AM	AM/PM
Stop Time	12:00 PM	AM/PM
Temp	71	F
Humidity	66	%
Ground	dry	dry/wet/snow
Air	dry	dry/rain/snow
Pressure	29.88	inHg
Wrap-around	35.6	mi

Datasheet - recorded 05/26/2019, page 1					
Northampton, MA					
Latitude			Longitude		
42	19	23.2	72	38	8.3
42° 19' 23.2" N			72° 38' 8.3" W		

	Station	Freq (KHz)	City	Lat			Long			Direction	Power Day (Watts)	Power Night (Watts)	Distance (mi)	Red Wire on M Coil, outbound	Measured Phase Shift (degrees)	Wrap-arounds	Initial Offset (degree)	Total Phase Shift (degree)	Phase Shift / Mile (degree)
1	WNNZ	640 AM	Westfield, MA	42	10	46	72	45	3	SSW	5,000	1,000	11.6	right	122	0	0	122	10.6
2	WGY	810 AM	Schenectady, NY	42	37	32	74	0	42	NW	50,000	50,000	73.3	right	41	2	0	761	10.4
3	WHMQ	1240 AM	Greenfield, MA	42	35	20	72	37	6	N	1,000	1,000	18.4	right	180	0	0	180	9.8
4	WHMP	1400 AM	Northampton, MA	42	19	36	72	39	26	NW	1,000	1,000	1.1	left	12	0	0	12	10.6
5	WTHE	1520 AM	Mineola, NY	40	44	45	73	37	28	SSW	1,000	0	120.5	right	24	3	0	1,104	9.2
																		average	10.1
																		standard deviation	0.6

Datasheet - recorded 05/26/2019, page 2

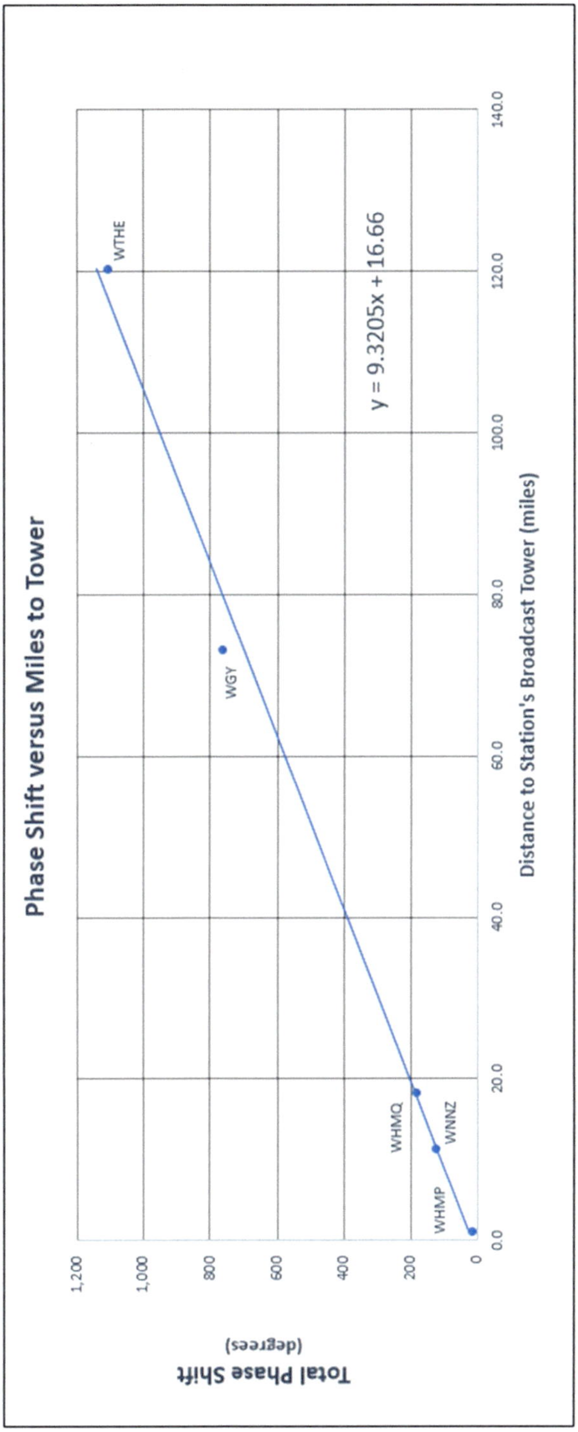

Start Time	3:00 PM	AM/PM
Stop Time	7:20 PM	AM/PM
Temp	73	F
Humidity	25	%
Ground	dry	dry/wet/snow
Air	dry	dry/rain/snow
Pressure	30.25	inHg
Wrap-around	33.6	mi

Datasheet - recorded 05/18/2019, page 1					
Northampton, MA					
Latitude			Longitude		
42	19	23.2	72	38	8.3
42° 19' 23.2'' N			72° 38' 8.3'' W		

	Station	Freq (KHz)	City	Lat			Long			Direction	Power Day (Watts)	Power Night (Watts)	Distance (mi)	Red Wire on M Coil, outbound	Measured Phase Shift (degrees)	Wrap-arounds	Initial Offset (degree)	Total Phase Shift (degree)	Phase Shift / Mile (degree)
1	WHYN	560 AM	Springfield, MA	42	11	37	72	41	0	SSW	5,000	1,000	9.3	right	101	0	0	101	10.9
2	WTAG	580 AM	Worcester, MA	42	20	13	71	49	13	E	5,000	5,000	41.7	right	129	1	0	489	11.7
3	WNNZ	640 AM	Westfield, MA	42	10	46	72	45	3	SSW	5,000	1,000	11.6	left	125	0	0	125	10.8
4	WACE	730 AM	Chicopee, MA	42	10	2	72	37	29	S	5,000	8	10.8	right	108	0	0	108	10.0
5	WGY	810 AM	Schenectady, NY	42	37	32	74	0	42	NW	50,000	50,000	73.3	right	117	2	0	837	11.4
6	WTIC	1080 AM	Hartford, CT	41	46	39	72	48	17	S	50,000	50,000	38.7	right	72	1	0	432	11.2
7	WHMQ	1240 AM	Greenfield, MA	42	35	20	72	37	6	N	1,000	1,000	18.4	right	167	0	0	167	9.1
8	WHMP	1400 AM	Northampton, MA	42	19	36	72	39	26	NW	1,000	1,000	1.1	right	12	0	0	12	10.6
																		average	10.7
																		standard deviation	0.8

Datasheet - recorded 05/18/2019, page 2

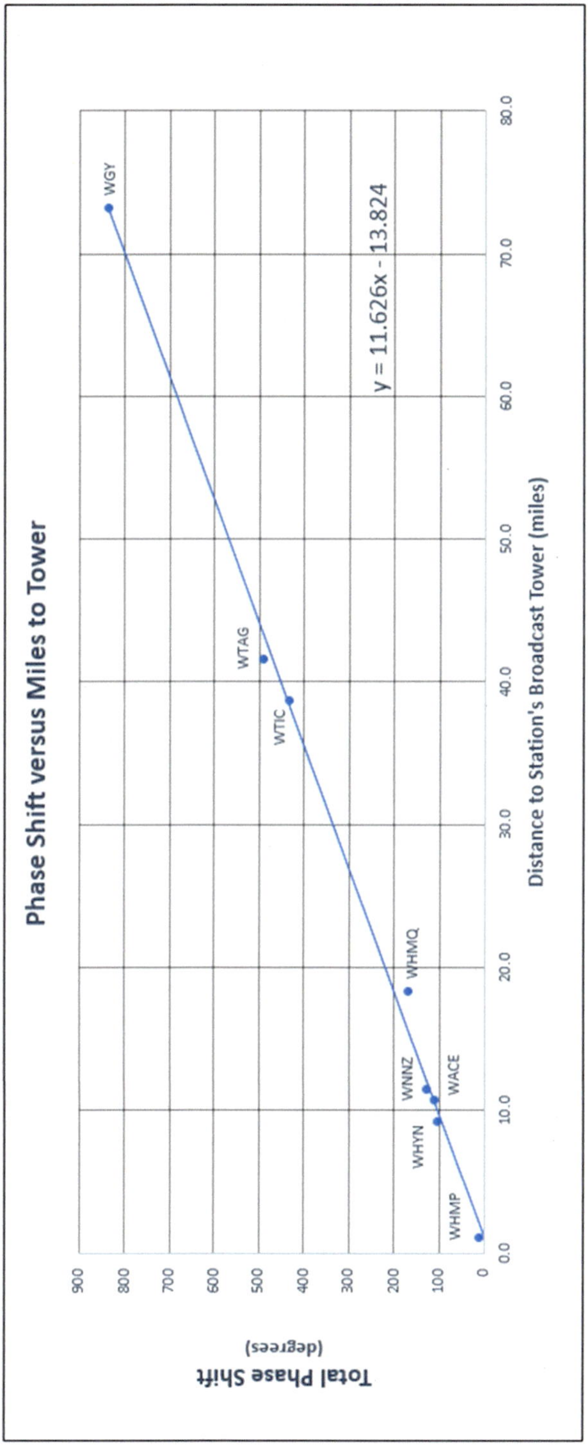

Start Time	8:30 AM	AM/PM
Stop Time	10:30 AM	AM/PM
Temp	48	F
Humidity	72	%
Ground	dry	dry/wet/snow
Air	~	dry/rain/snow
Pressure	29.81	inHg
Wrap-around	35.0	mi

Datasheet - recorded 05/15/2019, page 1					
Northampton, MA					
Latitude			Longitude		
42	19	23.2	72	38	8.3
42° 19' 23.2'' N			72° 38' 8.3'' W		

	Station	Freq (KHz)	City	Lat			Long			Direction	Power Day (Watts)	Power Night (Watts)	Distance (mi)	Red Wire on M Coil, outbound	Measured Phase Shift (degrees)	Wrap-arounds	Initial Offset (degree)	Total Phase Shift (degree)	Phase Shift / Mile (degree)
1	WHYN	560 AM	Springfield, MA	42	11	37	72	41	0	SSE	5,000	1,000	9.3	~	98	0	0	98	10.6
2	WTAG	580 AM	Worcester, MA	42	20	13	71	49	13	E	5,000	5,000	41.7	~	105	1	0	465	11.1
3	WNNZ	640 AM	Westfield, MA	42	10	46	72	45	3	SW	5,000	1,000	11.6	~	116	0	0	116	10.0
4	WACE	730 AM	Chicopee, MA	42	10	2	72	37	29	S	5,000	8	10.8	~	98	0	0	98	9.1
5	WGY	810 AM	Schenectady, NY	42	37	32	74	0	42	NW	50,000	50,000	73.3	~	63	2	0	783	10.7
6	WTIC	1080 AM	Hartford, CT	41	46	39	72	48	17	S	50,000	50,000	38.7	~	45	1	0	405	10.5
7	WARE	1250 AM	Ware, MA	42	14	43	72	12	27	E	5,000	2,500	22.6	~	198	0	0	198	8.8
8	WHMP	1400 AM	Northampton, MA	42	19	36	72	39	26	NW	1,000	1,000	1.1	~	13	0	0	13	11.5
																		average	10.3
																		standard deviation	0.9

Datasheet - recorded 05/15/2019, page 2

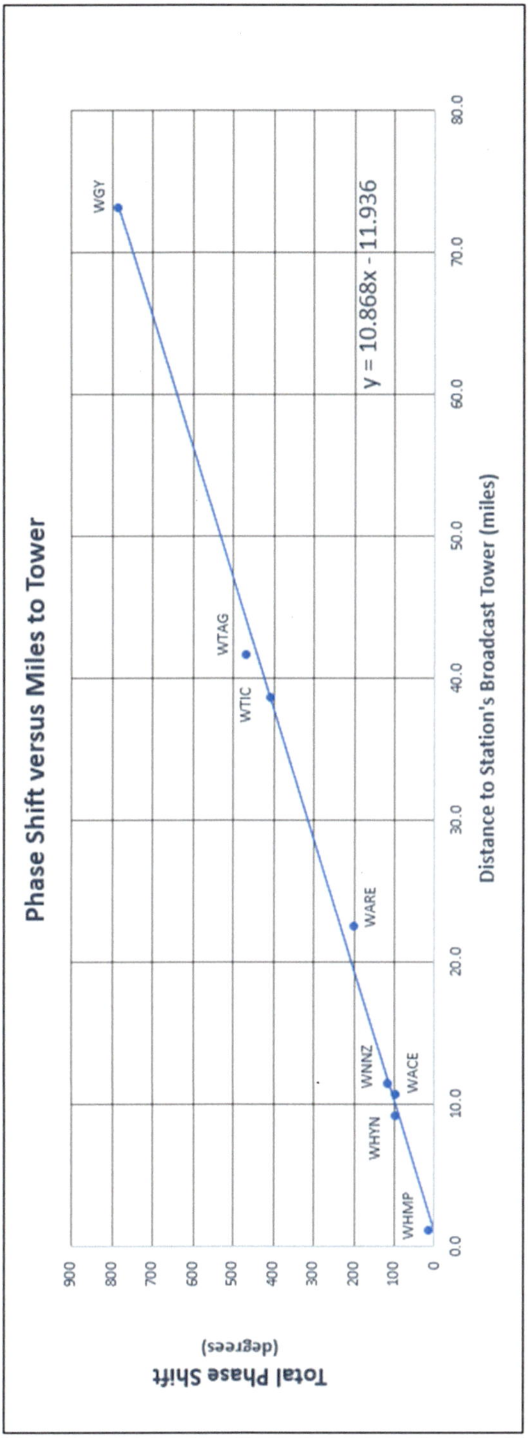

Start Time	1:20 PM	AM/PM
Stop Time	5:45 PM	AM/PM
Temp	41	F
Humidity	88	%
Ground	rain	dry/wet/snow
Air	~	dry/rain/snow
Pressure	30.03	inHg
Wrap-around	31.9	mi

Datasheet - recorded 05/12/2019, page 1					
Northampton, MA					
Latitude			Longitude		
42	19	23.2	72	38	8.3
42° 19' 23.2'' N			72° 38' 8.3'' W		

	Station	Freq (KHz)	City	Lat			Long			Direction	Power Day (Watts)	Power Night (Watts)	Distance (mi)	Red Wire on M Coil, outbound	Measured Phase Shift (degrees)	Wrap-arounds	Initial Offset (degree)	Total Phase Shift (degree)	Phase Shift / Mile (degree)
1	WHYN	560 AM	Springfield, MA	42	11	37	72	41	0	SSE	5,000	1,000	9.3	~	123	0	0	123	13.3
2	WTAG	580 AM	Worcester, MA	42	20	13	71	49	13	E	5,000	5,000	41.7	~	126	1	0	486	11.7
3	WNNZ	640 AM	Westfield, MA	42	10	46	72	45	3	SW	5,000	1,000	11.6	~	120	0	0	120	10.4
4	WACE	730 AM	Chicopee, MA	42	10	2	72	37	29	S	5,000	8	10.8	~	112	0	0	112	10.4
5	WVNE	760 AM	Leicester, MA	42	14	57	72	4	39	E	25,000	0	29.0	~	261	0	0	261	9.0
6	WGY	810 AM	Schenectady, NY	42	37	32	74	0	42	NW	50,000	50,000	73.3	~	99	2	0	819	11.2
7	WCRN	830 AM	Worcester, MA	42	14	50	71	55	50	E	50,000	50,000	36.5	~	74	1	0	434	11.9
8	WTIC	1080 AM	Hartford, CT	41	46	39	72	48	17	S	50,000	50,000	38.7	~	84	1	0	444	11.5
9	WARE	1250 AM	Ware, MA	42	14	43	72	12	27	E	5,000	2,500	22.6	~	286	0	0	286	12.7
10	WHLL	1450 AM	Springfield, MA	42	6	32	72	36	42	S	1,000	1,000	14.9	~	164	0	0	164	11.0
																		average	11.3
																		standard deviation	1.2

Datasheet - recorded 05/12/2019, page 2

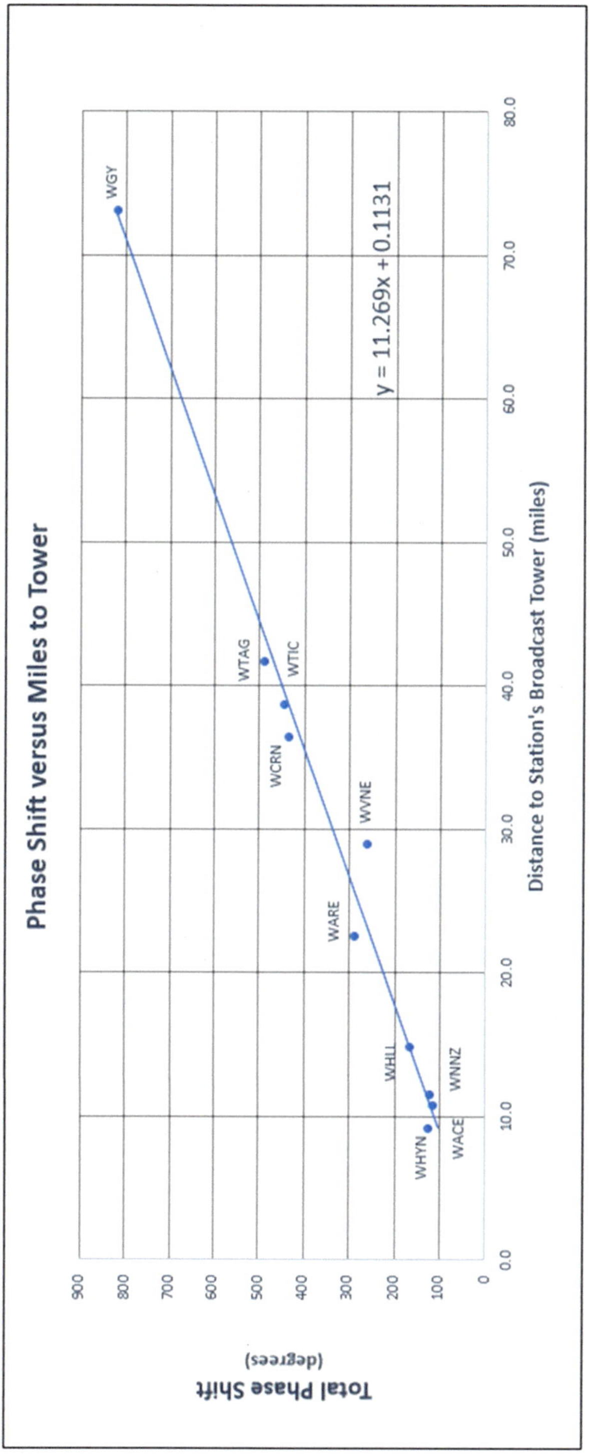

Appendix 6: Data – Mobile

Start Time	7:30 AM	AM/PM
Stop Time	1:00 PM	AM/PM
Temp	85	F
Humidity	51	%
Ground	dry	dry/wet/snow
Air	dry	dry/rain/snow
Pressure	29.94	inHg
Wrap-around	31.3	mi

Mobile

Readings taken on WTIC from five different locations

Datasheet - recorded 07/09/2020, page 1					
WTIC Broadcast Tower					
Latitude			Longitude		
41	46	39	72	48	17
41° 46' 39'' N			72° 48' 17'' W		

Station	Freq (KHz)	Location	Lat			Long			Direction	Power Day (Watts)	Power Night (Watts)	Distance (mi)	Red Wire on M Coil, outbound	Measured Phase Shift (degrees)	Wrap-arounds	Initial Offset (degree)	Total Phase Shift (degree)	Phase Shift / Mile (degree)
1 WTIC	1080 AM	Avon, CT	41	49	2	72	49	42	S	50,000	50,000	3.0	left	42	0	0	42	14.0
2 WTIC	1080 AM	Abigayles	41	50	51	72	48	36	S	50,000	50,000	4.8	left	62	0	0	62	12.8
3 WTIC	1080 AM	Simsburry, CT	41	52	33	72	48	1	S	50,000	50,000	6.8	left	75	0	0	75	11.0
4 WTIC	1080 AM	Holloways	41	54	16	72	47	8	S	50,000	50,000	8.8	left	85	0	0	85	9.6
5 WTIC	1080 AM	McLean Game Refuge	41	56	26	72	47	37	S	50,000	50,000	11.3	left	115	0	0	115	10.2
																	average	11.5
																	standard deviation	1.8

Datasheet - recorded 07/09/2020, page 2

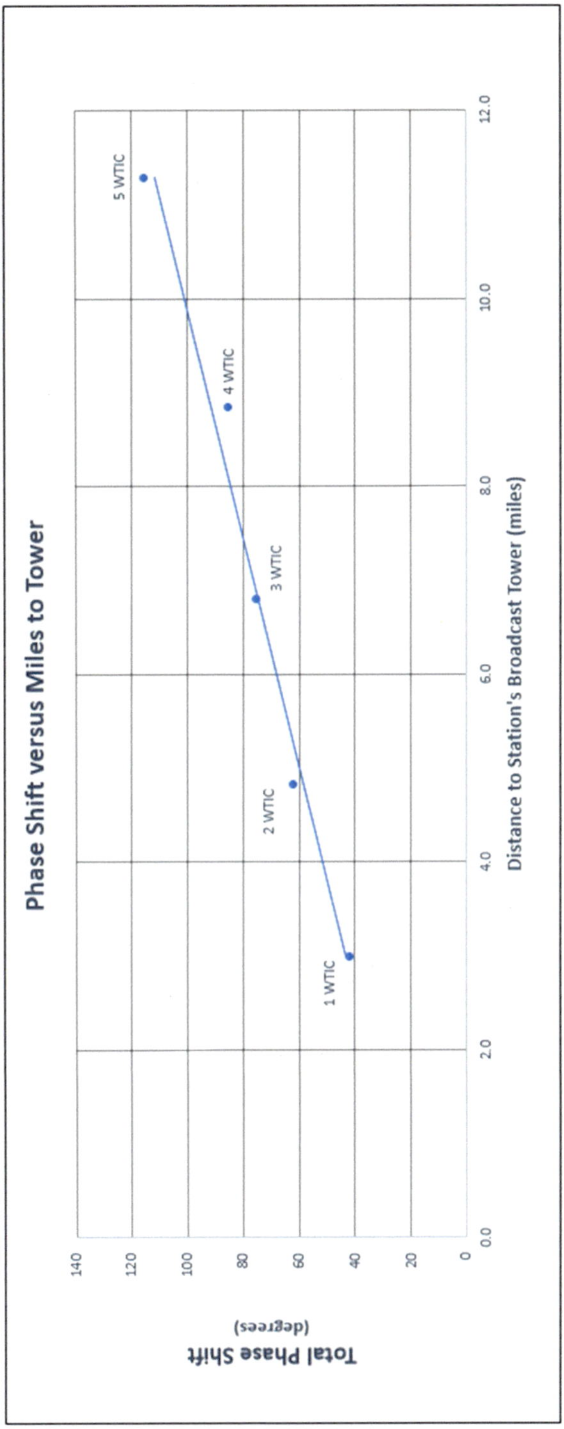

Start Time	9:00 AM	AM/PM
Stop Time	10:30 AM	AM/PM
Temp	59	F
Humidity	71	%
Ground	dry	dry/wet/snow
Air	dry	dry/rain/snow
Pressure	30.25	inHg
Wrap-around	31.1	mi

Mobile

Readings taken on WHMP from four different locations

Datasheet - recorded 05/23/2019, page 1					
WHMP Broadcast Tower					
Latitude			Longitude		
42	19	23.2	72	38	8.3
42° 19' 23.2'' N			72° 38' 8.3'' W		

	Station	Freq (KHz)	Location	Lat			Long			Direction	Power Day (Watts)	Power Night (Watts)	Distance (mi)	Red Wire on M Coil, outbound	Measured Phase Shift (degrees)	Wrap-arounds	Initial Offset (degree)	Total Phase Shift (degree)	Phase Shift / Mile (degree)
1	WHMP	1400 AM	Wallmart	42	20	33	72	38	39	S	1,000	1,000	1.4	~	13	0	0	13	9.3
2	WHMP	1400 AM	Double D Diner	42	22	20	72	38	6	S	1,000	1,000	3.4	~	31	0	0	31	9.1
3	WHMP	1400 AM	Log Homes	42	24	1	72	37	36	S	1,000	1,000	5.4	~	84	0	0	84	15.7
4	WHMP	1400 AM	Club Castaway	42	26	42	72	37	32	S	1,000	1,000	8.5	~	103	0	0	103	12.2
																		average	11.6
																		standard deviation	3.1

Datasheet - recorded 05/23/2019, page 2

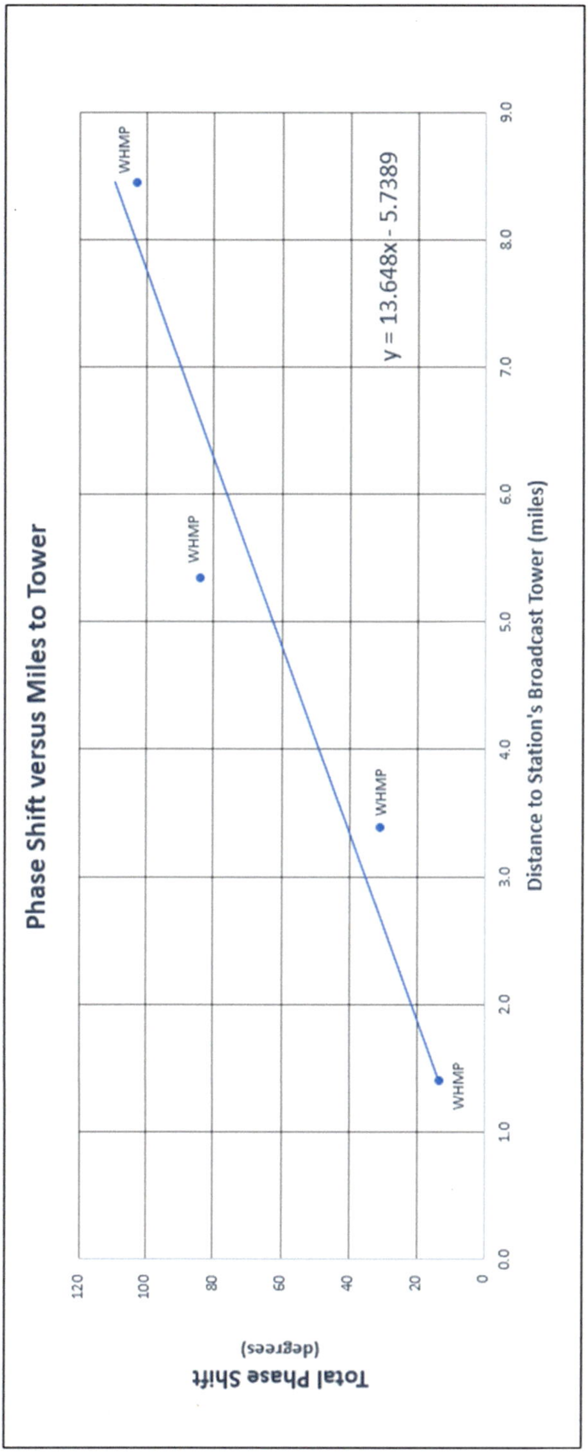

Appendix 7: Phase Shift Script

The code below can be inserted into a WaveForms (the application provided by Digilent to interface with the AD2 sampler) Script Window and run to calculate phase shift when two signals (M and E)are being displayed in the Scope1 window.

```
// M1 and M2 are the labels WaveForms uses
// for channels Math1 and Math2.
// It turns out that this script works fine
// for finding the phase shift between M1 and
// M2 except for when noise causes tiny (or large)
// local peaks, ... which is pretty much all the time
// unless filtering or averaging is turned on.
// Also, remember to set the value of the variable
// "base" to M1 or M2 before running to establish
// which channel is the reference, the magnetic.
// Duncan Crane 6/9/2019

function Phase(){
// Establish which two channels represent the M and the E after possible filtering
// through other channels.  For example the next two lines define channel "Math1"
// and "Math2" as the E and the M in the Scope1 window.  Then below M1 and M2
// are used to reference those channels.
//var d1 = Scope1.Math1.visibledata
//var d2 = Scope1.Math2.visibledata
var d1 = Scope1.Math6.visibledata
var d2 = Scope1.Math4.visibledata
var x1 = 0
var x2 = 0
var x3 = 0
var M1max1 = 0
var M1max2 = 0
var M2max1 = 0
var M2max2 = 0
var base

// set base to M1 or M2 here
base = "M2"
```

```
// find M1's first peak
for(var i = 0; i < (d1.length-2); i++){
x1 = d1[i]
x2 = d1[i+1]
x3 = d1[i+2]
if((x2>x1) && (x3<=x2)) {
M1max1=i+1
break;
}
if (i == d1.length) {
return "Error - no first peak found for Math1";
}
}

// find M1's second peak
for(var i = (M1max1+1); i < (d1.length-2); i++){
x1 = d1[i]
x2 = d1[i+1]
x3 = d1[i+2]
if((x2>x1) && (x3<=x2)) {
M1max2=i+1
break;
}
if (i == d1.length) {
return "Error - no second peak found for Math1";
}
}
// find M2's first peak
for(var i = 0; i < (d2.length-2); i++){
x1 = d2[i]
x2 = d2[i+1]
x3 = d2[i+2]
if((x2>x1) && (x3<=x2)) {
M2max1=i+1
break;
}
if (i == d2.length) {
return "Error - no first peak found for Math2";
}
}
// find M2's second peak
for(var i = (M2max1+1); i < (d2.length-2); i++){
x1 = d2[i]
```

```
x2 = d2[i+1]
x3 = d2[i+2]
if((x2>x1) && (x3<=x2)) {
M2max2=i+1
break;
}
if (i == d2.length) {
return "Error - no second peak found for Math2";
}
}
if (base == "M1") {
if (M2max1>=M1max1) {return ((M2max1-M1max1)/(M1max2-M1max1))*360}
else {return ((M2max2-M1max1)/(M1max2-M1max1))*360}
}
else if (base == "M2") {
if (M2max1>=M1max1) {return ((M1max2-M2max1)/(M2max2-M2max1))*360}
else {return ((M1max1-M2max1)/(M2max2-M2max1))*360}
}
else {return "Error - set vale of base to 'M1' or 'M2'"}
}
print(Phase())
```

Appendix 8: Authors

Duncan W. Crane

Duncan was born in Pittsfield, Massachusetts in 1947. He and his wife currently reside in western Massachusetts and are the parents of six children.

He served in the United States Air Force from 1966 to 1970 as a Ground Radio Repairman. He earned a Bachelor's of Science in Physics from Acadia University, Nova Scotia, Canada, and a Masters of Business Administration from the University of Massachusetts, Amherst.

Duncan worked as a computer programmer at a number of regional and national banks, and as a consultant data modeler and trainer at a number of financial and manufacturing companies throughout the United States. He retired in 2014.

His interests include guitar, sailing, meditation, science and family.

Robert Crane, Jr

Robert Crane, Jr with brother Duncan MacLaren Crane

Robert Crane Jr. was born in New York City in 1919 and died at the age of 97 in Southbury, Connecticut in 2017. He is the father of two children from his first marriage, and four from his second marriage.

Robert earned a Bachelor's of Science in Physics from Williams College, Williamstown, Massachusetts.

During World War II, Robert worked at the Radar Laboratory ("Rad Lab") at the Massachusetts Institute of Technology in Cambridge, Massachusetts. Then over the course of his career he worked at General Precision, Raytheon and Perkin Elmer.

Robert worked in the fields of radio, radar and optics. He had 19 patents to his credit. Three of his career accomplishments include:

> Origination of the Phase Measurement Interferometry (PMI) that greatly improved the precision of optical metrology and advanced the manufacturing and downsizing of computer chips.

Origination of the fine guidance sensor in the Hubble Telescope that ultimately made photographing deep space objects possible.

Origination of the laser illumination detection device that made it possible for all military personnel vehicles as well as satellites to know when they were being "lit up" with laser guided missiles. This was a difficult undertaking and thought impossible to do at the time because laser light is hard to differentiate from sun light.

Robert's brother Duncan, two years his junior, died during the invasion of Okinawa during WWII. This always caused Robert to question the value of his contributions to society when compared to his brother's sacrifice. The invention of the laser detection device that would benefit all military personnel in combat brought him deep personal satisfaction.

References

1. "A Student's Guide to Maxwell's Equations", Daniel Fleisch, Cambridge University Press, 2008
2. "Thinking about Science with David Hukins", thinking-about-science.com
3. "Lights two fields", Robert Crane, Proceedings of SPIE (International Society for Optics and Photonics), 10 Sept 2009
4. Digilent Inc., manufacturer of "Analog Discovery 2", a digital oscilloscope with dual channel analog-to-digital sampler
5. "Resonance and Coupled Circuits", BrainKart.com
6. "Light Dispersion and Spectrum", slide 3, Slideshare.net
7. "Dispersion", Jeffrey Mason, Perdue Physics, 2019, Docnet
8. "Ground Wave Propagation", Wikipedia, en.wikipedia.org
9. "Ground Wave Propagation", electronicsforu.com
10. "The Speed of Radio Waves and Its Importance in Some Applications (1950)", R. L. Smith-Rose, IEEE Xplore
11. "Causality Electromagnetic Induction and Gravitation", page 16, 2000, Oleg Dmitrovitch Jefimenko
12. "Circular Polarization", Wikipedia, en.wikipedia.org

Made in the USA
Monee, IL
05 November 2026

67953745R00062